3rd International Conference on Sustainable Energy Engineering (ICSEE 2019)

IOP Conference Series: Earth and Environmental Science
Volume 342

Shanghai, China
24 – 26 May 2019

ISBN: 978-1-7138-0140-5
ISSN: 1755-1307

Printed from e-media with permission by:

Curran Associates, Inc.
57 Morehouse Lane
Red Hook, NY 12571

Some format issues inherent in the e-media version may also appear in this print version.

Institute of Physics
Dirac House, Temple Back
Bristol BS1 6BE UK

Phone: 44 1 17 929 7481
Fax: 44 1 17 920 0979

techtracking@iop.org

Additional copies of this publication are available from:

Curran Associates, Inc.
57 Morehouse Lane
Red Hook, NY 12571 USA
Phone: 845-758-0400
Fax: 845-758-2633
Email: curran@proceedings.com
Web: www.proceedings.com

TABLE OF CONTENTS

POWER SYSTEM

RENEWABLE ENERGY

ENERGY SYSTEM

IOP Conf. Series: Earth and Environmental Science **342** (2019) 011001 doi:10.1088/1755-1315/342/1/011001

PREFACE

It is our great pleasure to welcome you to 2019 the 3rd International Conference on Sustainable Energy Engineering (ICSEE 2019) which will be held in Shanghai, China during May 24-26, 2019. ICSEE 2019 is dedicated to issues related to Sustainable Energy Engineering.

ICSEE is an annual International Conference on Sustainable Energy Engineering. ICSEE was started in 2017. After University of Western Australia, Perth, Australia on 2017, University of Technology Sydney, Sydney, Australia on 2018, Shanghai organize this third conference of a cycle of conferences on Sustainable Energy Engineering.

Explore the rapidly expanding roles of sustainable energy engineering in advanced energy generation and conservation. 2019 International Conference on Sustainable Energy Engineering (ICSEE 2019) brings together experts from across the energy value chain and addresses key topics on: Renewable Energy, Energy System and Power System.

Professors from UK, Canada and China are invited to deliver keynote speeches regarding latest information in their respective expertise areas. It will be a golden opportunity for the students, researchers and engineers to interact with the experts and specialists to get their advice or consultation on technical matters, sales and marketing strategies.

We would like to thank all the authors that submit their paper to the conference. We would like here to thank sincerely all our colleagues who extended their help in terms of organizing, reviewing, promoting, etc. for their continuous help and support and we promise you all that we will continue our efforts to make the joint ICCSS to be one of the best conferences.

Looking forward to meeting you always at our joint ICSEE conference.

Conference Chair
Prof. Zhixin Wang, Shanghai Jiaotong University, China
May 15, 2019

IOP Conf. Series: Earth and Environmental Science **342** (2019) 011001 doi:10.1088/1755-1315/342/1/011001

Statement of Peer Review

All papers published in this volume of IOP Conference Series: Earth and Environmental Science (EES) (Online ISSN: 1755-1315; Print ISSN: 1755-1307) have been peer reviewed through processes administered by the proceedings Editors. Reviews were conducted by expert referees to the professional and scientific standards expected of a proceedings journal published by IOP Publishing.

IOP Conf. Series: Earth and Environmental Science **342** (2019) 011002 doi:10.1088/1755-1315/342/1/011002

Conference Committee

Conference Chair

Zhixin Wang, Shanghai Jiaotong University, China
Dylan Lu, University of Technology Sydney, Australia

Program Chairs

Lin Chen, Chinese Academy of Sciences, China (Tohoku University, Japan)
Hee-Je Kim, Pusan National University, Korea
Chengke Zhou, Glasgow Caledonian University, UK

Technical Committees

Yahong Chen, Wuhan University, China
Renee Yi-Mond Yuan, Tamkang University, Taiwan
Xiangyun Chen, Inner Mongolia University of Technology, China
Hwan-Gyu Cho, Pusan National University, Korea
Muhammad Raza Ul Mustafa, Universiti Teknologi PETRONAS, Malaysia
Gopinath Chattopadhyay, Federation University, Australia
Nallusamy Nallusamy, SSN College of Engineering, India
Fengji Luo, The University of Sydney, Australia
Farhad Shahnia, Murdoch University, Australia
Ali Arefi, Murdoch University, Australia
Tsung- Mou Huang, National Sun Yat-Sen University, Taiwan
Muammer Ozgoren, Necmettin Erbakan University, Turkey
Iessa Sabbe Moosa, University of Buraimi, Oman
Radu Godina, New University of Lisbon, Portugal
Ghassan Fadil Lattif Al-Doori, University of Buraimi, Oman
Gopinath Chattopadhyay, Federation University, Australia
Ana Maria Diez Pascual, University of Alcala, Spain

IOP Conf. Series: Earth and Environmental Science **342** (2019) 011003 doi:10.1088/1755-1315/342/1/011003

First name	Surname
Hong	Cao
Yang	Sun
Xiangyun	Chen
Chao	Huang
Yuchun	Li
J J	Wang
J W	Lin
Zhili	Yao
JIN-SOO	BAK
Inho	Cho
Guo	Guo
Jafferi	Jamaludin
Mahmoud	Bady
Happy	Tshimbiluni
Shixiong	Qi
Haesung	Tak
Anselemi Babilas	Lukonge
Yahong	Chen
Zhichao	Liu
Yi	Luo
Chengke	Zhou
Yanchen	Liu
Jiaqi	Liao
Wenpei	Zhu
Chunyang	Gong
Yingqi	Yang
Mengshu	Shi

IOP Conf. Series: Earth and Environmental Science **342** (2019) 011003 doi:10.1088/1755-1315/342/1/011003

Institution

China Electric Power Research Institute, China

Dalian University of Technology, China

Inner Mongol University of Technology, China

Suzhou Nuclear Power Research Institute, China

State Grid Xinjiang Electric Power Company Electric Power Research Institute, China

Shanghai Jiao Tong University, China

Shanghai Jiao Tong University, China

Shanghai Jiao Tong University, China

Pusan National University , South Korea

Pusan National University , South Korea

Beijing Jiaotong University, China

Keio University, Japan

Islamic University of Medina, KSA

Durban University of Technology, South Africa

Xi'an Jiaotong University, China

Pusan National University, Korea

China University of Petroleum (East China), Qingdao, China

Wuhan University, China

Wuhan University, China

North China Electric Power University, China

Glasgow Caledonian University, UK

Hunan University, China

Wuhan University, China

Lanzhou University of Technology, China

Shanghai University of Electric Power, China

North China Electric Power University, Hebei, China

North China Electric Power University, Hebei, China

E-mail address
caohong@epri.sgcc.com.cn
1186435969@qq.com
lzg05006@163.com
huangchao2348@163.com
18361397336@163.com
wjj1252600254@163.com
sjtuljw@sjtu.edu.cn
yzl18217185269@163.com
xxximjsn@gmail.com
whdlsgh92@icloud.com
18117006@bjtu.edu.cn
jafferi@um.edu.my
mfbady@yahoo.com
happychristian42@gmail.com
qishixiong@stu.xjtu.edu.cn
tok33@pusan.ac.kr
lukongeab@gmail.com
yahongchen@foxmail.com
dengch@whu.edu.cn
lyphzh@163.com
c.zhou@gcu.ac.uk
491276835@qq.com
liaojiaqi@whu.edu.cn
5693665@qq.com
18508528566@163.com
15176261181@163.com
shimengshu1227@foxmail.com

IOP Conf. Series: Earth and Environmental Science **342** (2019) 011004 doi:10.1088/1755-1315/342/1/011004

Peer review statement

All papers published in this volume of *IOP Conference Series: Earth and Environmental Science* have been peer reviewed through processes administered by the proceedings Editors. Reviews were conducted by expert referees to the professional and scientific standards expected of a proceedings journal published by IOP Publishing.

IOP Conf. Series: Earth and Environmental Science **342** (2019) 012001 doi:10.1088/1755-1315/342/1/012001

Analysis of Fault Characteristics for UHVDC Hierarchical Connection to AC Grid

Hong Cao[1], Xingguo Wang[1], Min Xie[2]
1 State Key Laboratory for Security and Energy Saving, China Electric Power Research Institute, Beijing, 100192, China.
2 State Grid Anhui Electric Power Company, Hefei, 230021, China.

hongcao16@163.com

Abstract. UHVDC transmission project will adopt hierarchical connection mode at its receiving end with its high and low converters integrated to 500 kV and 1000 kV AC grids respectively on inverter side. The short-circuit fault at AC line of hierarchical connected UHVDC transmission system is a serious threat to the safe and stable operation of the system. In this paper, the equivalent model of UHVDC hierarchical connection to AC Grid is proposed. The dynamic phasor theory is used to analyse the fault characteristics of UHVDC system equivalent fundamental frequency current. When the AC fault occurs at the distal of the converter station, the UHVDC system equivalent power frequency current shows a decreasing trend; When the AC fault occurs at the proximal of the converter station, a high current occurs in the UHVDC system equivalent fundamental frequency current. RTDS model simulation verified that the theoretical analysis result is correct and established foundation to the subsequent protection impact study.

1. Introduction

Compared with the multi-infeed UHVDC transmission mode, UHVDC hierarchical connection to AC Grid means that the inverter side is connected to 1000kV and 500kV voltage level power grids respectively, which has the characteristics of low engineering cost and safe and stable operation of the power grid. It can effectively improve the voltage support capability of the receiving network. By directing the UHVDC power of the transmission to be properly distributed in different receiving circuits, the advantages of the transmission capacity of the two-stage power grid can be fully utilized[1-2].

At present, the research on UHVDC hierarchical connection to AC Grid mostly focuses on access mode [3-4], stability [5-6], control strategy [7], etc. Few literatures mention the fault characteristics of UHVDC hierarchical connection to AC Grid and their impact on protection. The short-circuit fault at AC line of hierarchical connected UHVDC transmission system is a serious threat to the safe and stable operation of the system. Therefore, it is urgent to study the fault characteristics of the UHVDC hierarchical connection to AC Grid. In this paper, the equivalent model of UHVDC hierarchical connection to AC Grid is proposed. The dynamic phasor theory is used to analyze the fault characteristics of UHVDC system equivalent fundamental frequency current. When the AC fault occurs at the distal of the converter station, the UHVDC system equivalent power frequency current shows a decreasing trend; When the AC fault occurs at the proximal of the converter station, a high current occurs in the UHVDC system equivalent fundamental frequency current. RTDS model

IOP Conf. Series: Earth and Environmental Science **342** (2019) 012001 doi:10.1088/1755-1315/342/1/012001

simulation verified that the theoretical analysis result is correct and lays the foundation for the subsequent protection impact study.

2. Equivalent Circuit

In severe cases, the AC side fault will cause large fluctuations in the AC bus, resulting in a commutation failure. At this time, the fluctuation of the direct current will increase, and the conduction of each valve will also change, so the current injected into the alternating current system will also change accordingly. Therefore, relative to pure AC system in power network of receiving end, the access of the UHVDC system is equivalent to the addition of a voltage-controlled current source $i_{dc} = f(U_{Bus})$, as shown in Figure 1. When the AC/DC hybrid system is in normal operation, i_{dc} behaves as a stable constant current source; In the event of a UHVDC system failure or disturbance that causes the commutation failure in the UHVDC system, Since the UHVDC system contains a large number of non-linear power electronics and rapid regulation of the control system in the UHVDC system, i_{dc} appears as a current source with complex harmonic components.

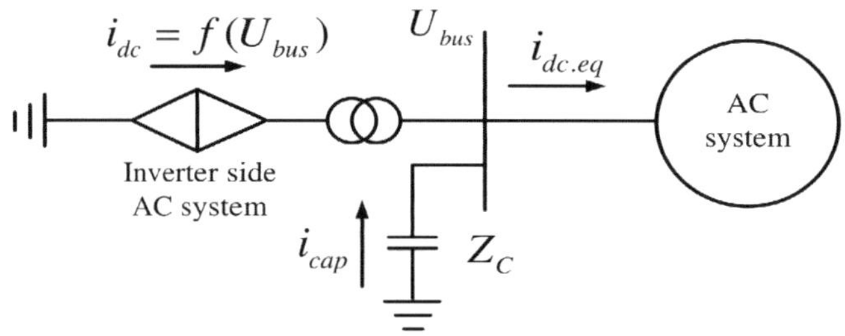

Figure 1. UHVDC system equivalent diagram

The equivalent power frequency current injected into the AC grid by the UHVDC system is:

$$I_{\varphi 1.eq} = I_{\varphi 1} + I_{cap1} \tag{1}$$

Where: $I_{\varphi 1}$ is the current power frequency component of the UHVDC system injected into the AC grid through the inverter during the transient process; I_{cap1} is the power frequency component of the reactive power compensation branch current I_{cap} injected into the AC grid by the UHVDC system.

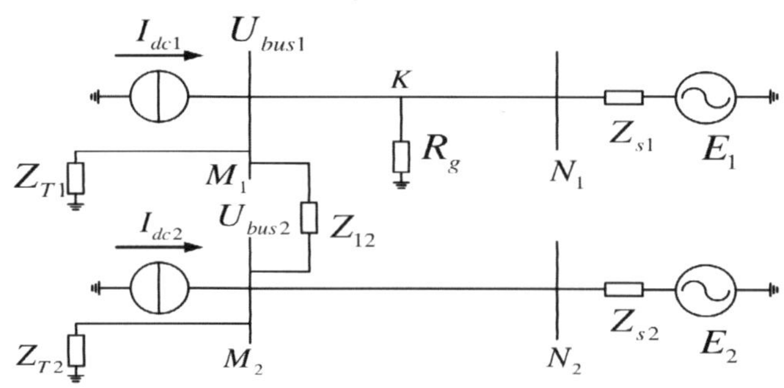

Figure 2. Schematic diagram of AC/DC hybrid system with UHVDC hierarchical connection to AC Grid

Schematic diagram of AC/DC hybrid system with UHVDC hierarchical connection to AC Grid is shown in Figure 2. Z_{c1} and Z_{c2} are the reactive power compensation branch impedance, Z_{12} is the connection impedance of the two-voltage AC grid, and I_{dc1} and I_{dc2} are the equivalent current of the high-end valve group and the equivalent current of the low-end valve group in UHVDC hierarchical connection system, R_g is the transition resistance, Z_{s1} and Z_{s2} are the equivalent impedance of the AC system, E_1 and E_2 are the equivalent internal potentials. When the AC grid

IOP Conf. Series: Earth and Environmental Science **342** (2019) 012001 doi:10.1088/1755-1315/342/1/012001

failure, due to the nonlinear characteristics of the converter, the UHVDC system equivalent currents is dynamically changing, which can be considered as dynamic phasor with time-frequency domain characteristics.

3. Fault Characteristics of UHVDC System Equivalent Fundamental Frequency Current

In this paper, the model of the UHVDC system is established by dynamic phasor theory, and the change of UHVDC system equivalent fundamental frequency current after AC fault is analytically calculated.

Based on the time-varying Fourier transform, the waveform of a signal $x(\tau)$ in the time domain can be expanded into a time-varying Fourier series in any interval $\tau \in (t-T, t)$:

$$x(\tau) = \sum_{k=-\infty}^{\infty} X_k(t) e^{jk\omega\tau} \tag{2}$$

Where: ω=2π/T, $X_k(t)$ is the k-timed variable Fourier coefficient, and the kth coefficient can be obtained by:

$$X_k(t) = \frac{1}{T}\int_{t-T}^{t} x(\tau) e^{-jk\omega\tau} d\tau = \langle x \rangle_k (t) \tag{3}$$

Where $X_k(t)$ is a function of time. When the window under study (width T) moves along the time axis on the waveform x(τ), the phasor $X_k(t)$ changes.

Under normal operating conditions, the converter valves are sequentially turned on in sequence, and the relationship between the voltage and current on both sides of the AC and UHVDC can be expressed as:

$$\begin{cases} u_{di} = u_{ai}S_{ua} + u_{bi}S_{ub} + u_{ci}S_{uc} \\ i_{ai} = i_{di}S_{ia} \\ i_{bi} = i_{di}S_{ib} \\ i_{ci} = i_{di}S_{ic} \end{cases} \tag{4}$$

Where: subscript "i" means inverter, u_{di} is inverter side UHVDC voltage, i_{di} is inverter side UHVDC current; S_{ua}, S_{ub} and S_{uc} are abc three-phase voltage switching functions respectively; S_{ia}, S_{ib} and S_{ic} are respectively Three-phase current switching function. The switching function is to treat the converter valve as an electronic switch. The square wave has an amplitude of 1 during the conduction of the converter valve and a value of 0 during the off period.

It can be seen from equations (4) and (5) that the dynamic phasor form of the alternating current on the inverter side is:

$$I_{\varphi(k)} = \sum_m I_{dc(k-m)} S_{i\varphi(m)} \tag{5}$$

Where: k and m represent the order of the corresponding dynamic phasors, respectively, and φ is the three phases a, b, and c. Equation (6) is a dynamic phasor model for the current injected into the AC grid by the converter. Since the existing protection principle is mainly based on the power frequency component, this paper analyzes the variation of the equivalent current power frequency component injected into the AC system through the inverter during the transient process, that is, taking k= in equation (6). 1, get:

$$I_{\varphi 1} = \sum_m I_{dc(1-m)} S_{i\varphi(m)} \tag{6}$$

Substituting equation (7) into equation (1):

$$I_{dc.eq} = \sum_m I_{dc(1-m)} S_{i\varphi(m)} + I_{cap} \tag{7}$$

It can be known from equation (8) that the equivalent fundamental frequency current of the UHVDC system is related to the three-phase current switching function after the failure of the AC system. The switching function is directly affected by the commutation bus voltage. When the AC fault occurs at the distal of the converter station, the voltage of the commutation bus is less affected.

IOP Conf. Series: Earth and Environmental Science **342** (2019) 012001 doi:10.1088/1755-1315/342/1/012001

Due to the influence of the control system, the equivalent fundamental frequency current of the UHVDC system decreases. When the AC fault occurs at the proximal of the converter station, the bus voltage is greatly affected, and the UHVDC system equivalent fundamental frequency current has a high current.

4. Simulation

Based on parameters of the domestic ±1100 kV UHVDC project with hierarchical connection mode, as shown in Fig.3, a simulation model is built on RTDS. The UHVDC inverter side is layered into a 500 kV/1000 kV AC hierarchical power grid. The UHVDC rated voltage is ±1100kV and the UHVDC rated current is 5.45KA.

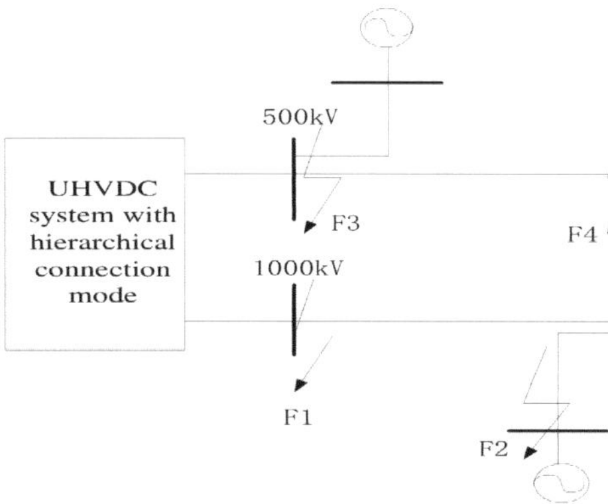

Figure 3. Structure of UHVDC system with hierarchical connection mode

When the A-phase fault is happened at the proximal of the 1000kV AC grid, the fundamental frequency current amplitudes at the 500kV commutating bus and the 1000kV commutating bus are shown in Figure 4 and Figure 5 respectively.

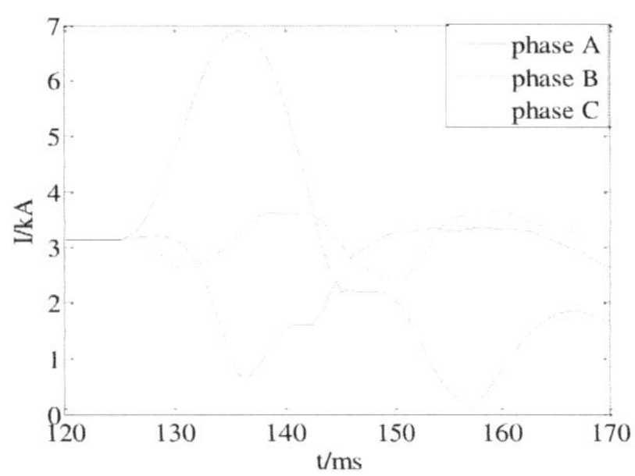

Figure 4. Fundamental frequency current amplitude waveform at 1000kV commutation bus

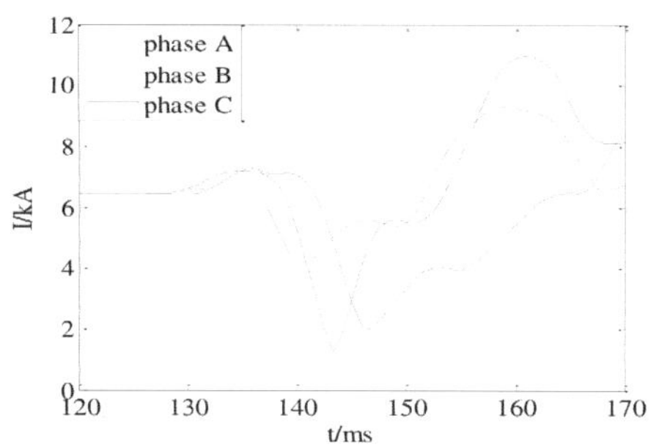

Figure 5. Fundamental frequency current amplitude waveform at 500kV commutation bus

When the A-phase fault is happened at the distal of the 1000kV AC grid, the fundamental frequency current amplitudes at the 500kV commutating bus and the 1000kV commutating bus are shown in Figure 6 and Figure 7 respectively.

It can be seen from Figure 4-7, when the distance between fault location and the commutation bus is lager, the UHVDC system equivalent fundamental frequency current is smaller, and it shows a downward trend in the initial stage of the fault. Simulation results of 1000kV distal and proximal faults are similar to those of 500kV distal and proximal faults, which are not given in this paper due to the limitation of length.

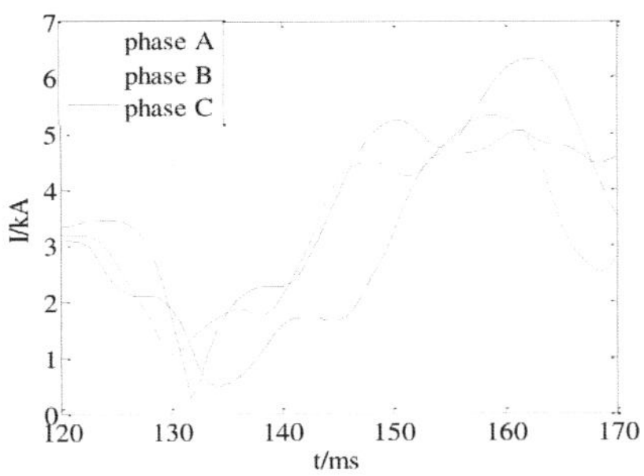

Figure 6. Fundamental frequency current amplitude waveform at 1000kV commutation bus

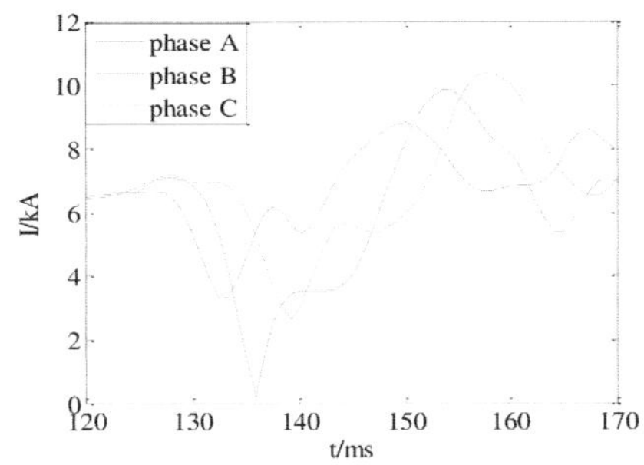

Figure 7. Fundamental frequency current amplitude waveform at 500kV commutation bus

5. Conclusion

In this paper, the equivalent model of UHVDC hierarchical connection to AC Grid is proposed. The dynamic phasor theory is used to analyse the fault characteristics of UHVDC system equivalent fundamental frequency current. When the AC fault occurs at the distal of the converter station, the UHVDC system equivalent power frequency current shows a decreasing trend; When the AC fault occurs at the proximal of the converter station, a high current occurs in the UHVDC system equivalent fundamental frequency current. RTDS Simulation tests verify the validity of the theoretical analysis. It is known from the equivalent fundamental frequency current characteristics of the UHVDC system that the distance protection performance of AC system is seriously affected, and the adaptability of the distance protection needs further study.

6. References

[1] Zhenya Liu, Xiaohui Qin, liang Zhao, et al. Study on the Application of UHVDC Hierarchical Connection Mode to Multi-infeed HVDC System. J. *Proceedings of the CSEE,* 2013, 33(10):1-7.

[2] Yi Tang, Bin Chen, Qi Wang, et al. Analysis on Reactive Power and Voltage Coupling Characteristics of Hybrid System for UHVDC Hierarchical Connection to AC Grid. J. *Power System Technology,* 2016, 40(4):1005-1011.

[3] Yu Shen, Yongxin Xiong, Wei Yao, et al. Comprehensive Evaluation for Receiving End Connection Scheme of ±1100kV UHVDC Power transmission. J. *Electric Power Automation Equipment,* 2016, 40(4):1005-1011.

[4] Sun W, Rong F, Zhou Z, et al. Analysis on the operating characteristic of UHVDC hierarchical connection mode to AC system. J. *Proceedings of the CSEE,* 2013, 33(10):1-7.

[5] Wang C, Zhu L, Chen B, et al. Analysis on voltage stability of hybrid system with UHVDC hierarchical connection to AC grid. *Power Electronics & Motion Control Conference,* 2016.

[6] Jia N, Qi W, Yi T, et al. Study on the characteristics of AC/DC hybrid system under UHVDC hierarchical connection mode. *Power Electronics & Motion Control Conference,* 2016.

[7] Yongxin Xiong, Yu Shen, Wei Yao, et al. Improved Additional Power Coordinated Control of ±1100kV UHVDC System With Hierarchical Connection Mode. J. *Power System Technology,* 2017(11):3448-3456.

Acknowledgments

This work is supported by Science and Technology Project of State Grid Corporation of China: Research on relay protection optimization strategy of ±1100kV DC/ 1000kV AC hybrid power system (SGAH0000TKJS1800050).

IOP Conf. Series: Earth and Environmental Science **342** (2019) 012002 doi:10.1088/1755-1315/342/1/012002

Research on International Power Network Construction in Central-Asia

Juan Li[1], Yang Sun[2], Hong-lian Zhou[1], Nan Yang[2], Ling-shuang Liu[1], Er-biao Zhou[1], Zhong-zheng Li[1] and Tie-jiang Yuan[2]

1.State Grid Xinjiang Electric Power Research Institute, Urumqi 830002, China
2. School of Electrical Engineering, Dalian University of Technology, Dalian 116024, China

1186435969@qq.com

Abstract. As part of the Belt and Road Initiative, China promises to provide its abandoned renewable energies in Xinjiang for solving the power shortage problem in Central-Asia countries. This brings a great chance for energy investment and cooperation in power grid construction between China and Central-Asia countries. This paper analyses the power structure and energy institutions in 6 Central-Asia countries and predicts its energy potential and power requirement based on its GDP growth rate. A power grid connection plan is further proposed based on the predictions. This power grid construction would benefit all the countries involved and boost area economic growth for years.

1. Introduction

A green and clean electricity production method has been widely recognised as an urgent requirement internationally. However, in some inland developing countries, due to insufficient financial and technical support, their power production is still dominated by fossil fuels power plants [1].
Moreover, due to the relatively high economic growth rate in Central Asia, power shortage has become a serious problem for them. However, without financial and technical support, fossil-fuel power plants would still be their first choice in the future energy development. This would make the global carbon emission tasks much more difficult to finish and risk damaging the environment [2]. One of the typical area mentioned above is Central Asia area, including Pakistan, Kazakhstan, Tajikistan, Kyrgyzstan, Turkmenistan, Uzbekistan and Xinjiang, China.

As stated by President Xi Jinping in the 2015 New York United Nations Development Summit(UNDS), China proposes the construction of global energy network and dedicates in promoting a green and clean electricity production method. Xinjiang Province is one of the key power bases in China. The renewable energies have developed rapidly in recent years, reaching 22 MWh[1]. It also has the most advanced High-voltage Direct Current (HVDC) power transmission system and its corresponding power grid construction.

HVDC transmission is superior to AC transmission in international power transmission due to its improved stability and power capacity. By using HVDC , the reliability and transmission flexibility of the transmission system is improved.

As proposed in Belt and Road Initiative [1]–[3], China is willing to share its abundant electricity and advanced power technologies to other developing countries, seeking mutual interests and greater prosperities. The construction of a cross-country power network based on HVDC technologies can reduce the use of fossil-fuels, optimize the unbalanced source-load distribution and promote the economic growth both in Central-Asia and China.

This paper researches the current power system status in 6 Central-Asia countries. A prediction of their future energy requirement is carried based on their economic growth rate. Based on the prediction, the construction plan of global power network is regulated.

2. Current power system status in Central-Asia

The current power structure of 6 Central-Asia countries is discussed in this section.

2.1. Kazakhstan

According to the stats from Kazakhstan energy department, there are currently 71 power plants in operation, producing 19000 MW power. 83.3% of them come from thermal power plants, 4.8% from gas-turbine, 11.9% from hydropower stations. 70% of the power generation relies on coal, 15% on natural gas and heavy oil [4]–[6].

There are lots of wind and photovoltaic resources available in South and Southwest Kazakhstan. Although renewable energies have been promoted by the government for years, only 1500kW wind farms are currently under construction.

The major problem in Kazakhstan's power structure is concluded as:

- High industrial power consumption
- High carbon emission
- Heavy air pollution in the urban area
- Power shortage in the remote area

With the rapid economic development, the annual power demand increase in Kazakhstan is 5%-6%. If the economic growth rate kept constant over the next few years. Annual power increase rate is estimated to be 7.2% by 2020. Significant power shortage will be a major problem in South Kazakhstan.

Due to the relatively low infrastructure investment and technological support, no significant improvement can be founded in Kazakhstan's future development plans. Kazakhstan government has declared a national law for encouraging the development of renewable energies in 2011. However, even by the goal, renewable energies will only take around 1% of Kazakhstan's total power requirement.

2.2. Kyrgyzstan

Only 18 power plants are currently operating in Kyrgyzstan, producing 3748 MW power. 80.8% of them come from 16 hydropower plants, while the rest come from 2 thermal stations [7]–[9]. While the hydropower is a clean and green renewable energy, its dominated role in Kyrgyzstan does cause a series of the problem including: seasonal power variation, lack of resources and insufficient power regulation ability.

In opposite to Kazakhstan, Kyrgyzstan government tends to promote the development of renewable energies continuously. In the foreseeable future, the Kyrgyzstan government has proposed over 20 policies for encouraging renewable energies investment including land tax reduction.

An annual power consumption of each resident in Kyrgyzstan has exceeded 2400 kWh due to its fast industrial and agriculture development. However, due to its relatively low power supply (3746MW), continuous increase of electricity price (Resident electricity price: $0.043 to $0.1/kWh) has been conducted by the government to encourage energy-saving, significantly limiting the economic development. Estimated annual power requirement is 30 TWh, the maximum power load is 6000 MW by 2020, if the growth rate kept constant.

2.3. Uzbekistan, Turkmenistan

The power stations installed in Uzbekistan has reached more than 12400MW. There are 45 power plants in Uzbekistan, containing 13 thermal power plants (88.2%) and 32 hydropower plants (11.8%). As the largest national power company, Uzbekenergo has 16 power stations with 12040MW installed capacities. An only limited number of photovoltaic projects are declared by Uzbekistan government. 3

photovoltaic power stations are planned by 2020 in Samarkand, Namangan, and Suhan River. The total power capacities are estimated by 0.1 MWh and the ETA is around 2020[10], [11].

The install power capacities in Turkmenistan consists of 2857 MW gas-based thermal power plants, 1.2MW hydropower plants, 1592.7MW coal-based thermal power plants, and 400MW oil-based power plants. After the agreement with the Generic Electric company(USA) in 2006, the install capacities are expected to be increased by 50% in 2020[12]–[14].

The current power status in Uzbekistan and Turkmenistan are better than the previous countries. They are expected to be not only self-supported but also export energy to the other countries. By 2020, Uzbekistan is expected to have a total power requirement of 90 TWh and the maximum power load is 16000 MW. Turkmenistan, on the other hand, should have 30 TWh and 5500 MW. Itself can export 6 TWh to other countries, making it the most important power hub in Central-Asia. Pakistan should have 12.09 TWh total power requirement by 2015.

2.4. Tajikistan
Similar to Kyrgyzstan, the power structure in Tajikistan is also hydropower dominated. With over 98% of its 5400MW power from hydropower station, it shares a similar structure problem to Kyrgyzstan. 2 thermal power plants are used as the backup supply, generating only 318MW power to the grid.

Unlike Kyrgyzstan, Tajikistan doesn't have abundant wind resources. Currently, it focuses on the development of photovoltaic power and has a 160KW experiment system in Gyakov National Hospital and Institute of Maternal Science in Dushanbe. Although a nuclear power trade agreement has been signed with Russia in 2012, no nuclear power plant has been planned in the foreseeable future[15] .

The power prediction in Tajikistan is kind of unique in Central-Asia. Even its annual power production can theoretically meet its annual power requirement. Although the abundant water resources in the rainy season can not only provide enough power in summer but also have enough power for export, Tajikistan still needs to import power from other countries in winter seasons as shown in Fig.1. Therefore, Tajikistan focuses more on balancing the seasonal power variation. This makes the power network an excellent choice for Tajikistan.

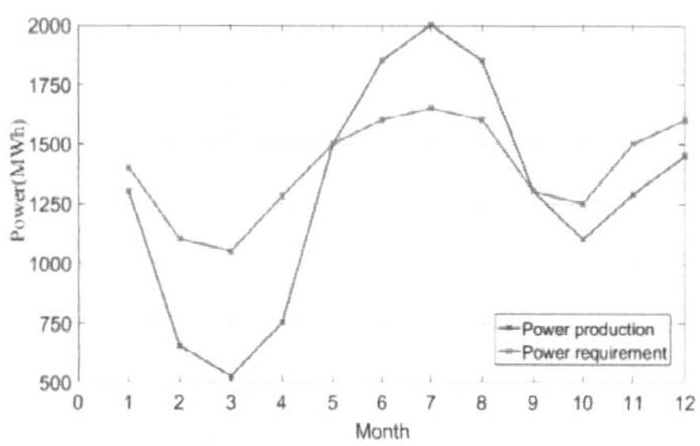

Figure 1. Power status in Tajikistan

Figure 2. Centre-Asia Super Power grid

2.5. Pakistan
By June 2015, the total install capacities in Pakistan is 24823 MW, increased by 6.3% compared to last year. Thermal plants take more than half its total power supply (67.74%), leading hydropower (28.66%) and a small number of other energies (3.6% nuclear and wind power)[16].

With the economy develops faster and faster in Pakistan, the power shortage problem is getting worse. It has become an important factor restricting economic development. According to information from Pakistan, the largest electricity demand in Pakistan is 13490 MW, while the capacity to generate electricity is 10790 MW and the electricity gap is 2700MW. The lack of power generation capacity is the main cause of the power crisis in Pakistan.

2.6. Summary of Power connection status in Central-Asia
As stated in the previous sections, several international power transmissions lines have been constructed for years as shown in Fig.2. To conclude, the Central-Asian super grid is a North-South oriented, tree-structured, 500KV single-loop power grid. However, due to the insufficient financial and technical support, this giant Central-Asian power network is currently facing the following

challenges: fragile power structure, high transmission losses, low transmission capabilities, ageing power devices, out-of-date control strategies, and insufficient covering areas.

With the increasing power demand, the complexity and the capacities of this power grid would increase exponentially, requiring an upgrade on both the transmission hardware (power devices) and transmission software (Control system). The cooperation between China and Central-Asian countries can solve the power dilemma in Central-Asia and significantly boost the economic development in this area.

3. International power network construction plans

The average distances between the capitals of Central-Asia countries and Urumqi, China are over 1000km. By considering the development of future wind-photovoltaic and hydropower plants, a ±500KV HVDC transmission line is more appropriate from both economic and technical point of view. When the two places are adjacent, the back-to-back DC transmission mode is selected, and the advantages are more prominent. There is no mainstream line, the DC loss is small, and the investment cost is low. It cooperates well with the current HVDC transmission system and satisfies the power demand in Central-Asia for the next decades. The typical parameters for an HVDC transmission line are shown in Table.1.

Table 1. The arrangement of channels.

Item	Value
Voltage Level	±500KV
Structure	4*300
Phase-Phase distance	59.4
R_0(Resistance)	0.02625
R_0^*(Resistance unit value)	$1.052*10^{-5}$
X_0(Reactance)	0.284
X_0^*(Reactance unit value)	$9.775*10^{-3}$
B_0(Conductance)	$3.910**10^{-6}$
B_0^*(Conductance unit value)	$9.775*10^{-3}$
Diameter	42.0

A prediction of Central-Asian power requirement is conducted by using import/export power data and economic growth rate as shown in Equ.1.

$$\begin{cases} R_W = K \times R \\ W_n = W_0 (1 + R_W)^n \end{cases} \tag{1}$$

Where R_w is the growth rate of power requirement, K is the elastic coefficient, R is the Growth rate of GDP, W_n is the predicted power requirement, W_0 is the current power requirement. The import and export power anticipation are shown in Table.2 (+ represents export power, - represents imported electricity) and its supply ability is shown in Table.3

Table 2. Import/Export energy estimation (TWh).

Country	2017	2020	2025	2030
China(Xinjiang)	-43490	-54830	-80550	-90301
Kazakhstan	531	615.6	713.5	827.3
Kyrgyzstan	-780	-560.8	-356.2	-124.5
Uzbekistan	-2830	-2360.3	-1896.2	-1328.2
Turkmenistan	600	705.6	825.3	948.5
Tajikistan	-210	-142.6	85.3	125.2
Pakistan	880	2320	3102	4234

Based on this prediction, the cross-country power grid connecting plan is proposed as shown in Fig.3. A ±500KV will be constructed between Xinjiang and Central-Asia, preferably to Kazakhstan since it is the major load Central, which would enable annual power transmission capability to reach 3000 MWh. The second transmission line (back to back transmission) with 1000 MWh transmission ability should be built between Kyrgyzstan and Xinjiang. The third one being the China-Pakistan transmission line with 6000 MWh capability which would cover Pakistan's power requirement until 2030.

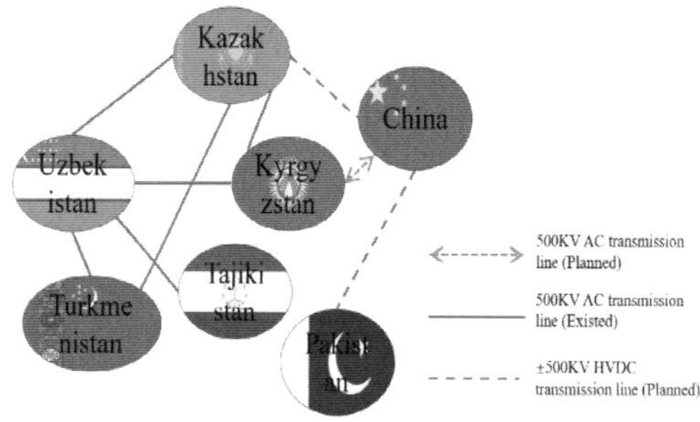

Figure 3. Power Grid construction plans

Table 3. Power Transmission ability (Unit: TWh).

Year	Power storage		Power transmission Ability	
	Central-Asia	Pakistan	Central-Asia	Pakistan
2017	-2464900	880000	/	/
2020	-1742500	2320000	-1642	2900
2025	-628300	3102000	-1200	4140
2030	448300	4234000	640	5369

4. Conclusion

A forthcoming power shortage can be expected in Centre-Asia due to the lack of financial and technical support in power construction. A China-Central Asian super power grid, which transfers the abandoned renewable energies internationally, would totally solve the power shortage problem in Centre-Asia, creating numerous chances in industrial and infrastructure development for both side. In this paper, the amount of power shortage is calculated based on the GDP. The construction plan of the China-Central Asian super power grid is regulated based on the calculation considering their economic benefits and technical issues. This proposed plan reduces the use of fossil-fuels and optimizes the unbalanced source-load distribution with 2 HVDC between Kazakhstan and Pakistan, and one DC back to back power grid in Tajikistan. Its detail construction plan (including connecting position and transmission control strategies) requires more information on the specific power transmission grid and will be studied in the future.

5. References

[1] Yigai Tang. Petroleum Geopolitics in Central Asia and Consumption of a Main Energy Security Corridor in Xinjiang. *Social Sciences in Xinjiang*, vol. 2007, no. 1, pp. 17-21+115, 2007.

[2] F. Indeo. The Impact of the Belt and Road Initiative on Central Asia: Building New Relations in a Reshaped Geopolitical Scenario. in *China's Belt and Road Initiative: Changing the Rules of Globalization*, W. Zhang, I. Alon, and C. Lattemann, Eds. Cham: Springer International Publishing, 2018, pp. 135–153.

[3] I. Qoraboyev and K. Moldashev.The Belt and Road Initiative and Comprehensive Regionalism in Central Asia. in *Rethinking the Silk Road: China's Belt and Road Initiative and Emerging Eurasian Relations*, M. Mayer, Ed. Singapore: Springer Singapore, 2018, pp. 115–130.

[4] M. Karatayev and M. L. Clarke. A review of current energy systems and green energy potential in Kazakhstan. *Renewable and Sustainable Energy Reviews*, vol. 55, pp. 491–504, 2016.

[5] M. Karatayev and M. L. Clarke. Current Energy Resources in Kazakhstan and the Future Potential of Renewables: A Review. *Energy Procedia*, vol. 59, pp. 97–104, 2014.
[6] M. Karatayev, S. Hall, Y. Kalyuzhnova, and M. L. Clarke. Renewable energy technology uptake in Kazakhstan: Policy drivers and barriers in a transitional economy. *Renewable and Sustainable Energy Reviews*, vol. 66, pp. 120–136, 2016.
[7] "Electric Energy consumption Per Capita in Kyrgyzstan | Helgi Library." [Online]. Available: https://www.helgilibrary.com/indicators/electric-power-consumption-per-capita/kyrgyzstan/. [Accessed: 17-Dec-2018].
[8] F. A. Hill, K. C. Minbaeva, M. A. Wilson, and R. Satylkanov. Hydrologic Controls and Water Vulnerabilities in the Naryn River Basin, Kyrgyzstan: A Socio-Hydro Case Study of Water Stressors in Central Asia. *Water*, vol. 9, no. 5, 2017.
[9] L. Guo, H. Zhou, Z. Xia, and F. Huang. Evolution, opportunity and challenges of transboundary water and energy problems in Central Asia. *SpringerPlus*, vol. 5, no. 1, p. 1918, 2016.
[10] E. Y. Rakhimov, S. E. Sadullaeva, Y. G. Kolomiets, K. K. Tashmatov, and N. O. Usmonov, "Analysis of the solar energy potential of the Republic of Uzbekistan *Applied Solar Energy*, vol. 53, no. 4, pp. 344–346, 2017.
[11] R. A. Zakhidov and S. L. Lutpullayev. Global trends in alternative energies and problems in Uzbekistan for the development of renewable energy sources. *Applied Solar Energy*, vol. 51, no. 1, pp. 50–61, 2015.
[12] Ning Zhang, *Central Asian energy status and game playing between super powers*. Changchun, China: Changchun Publisher, 2009.
[13] Yu Zhen, *Cooperation and Competition in Central-Asia: China, USA, Russia*. Beijing: Social Science literature Publisher, 2007.
[14] F. Costa Buranelli. One Belt, One Road and Central Asia: Challenges and Opportunities. in *The Belt & Road Initiative in the Global Arena: Chinese and European Perspectives*, Y. Cheng, L. Song, and L. Huang, Eds. Singapore: Springer Singapore, 2018, pp. 207–230.
[15] Shvedov GV, Chorshanbiev SR, Vaskov AG. Analysis and Evaluation of Potential of Renewable Energy Resources of Republic of Tajikistan.: *IEEE*, 2018: 26-33.
[16] Ali SM, Rashid A, Mehmood CA, et al. A survey of energy potential in Pakistan for smart grid implementation: *IEEE*, 2014: 443-47.

Acknowledgments
This work was supported by the science and technology program of Xinjiang Electric power CO.LTD(SGXJJY00SJJS1800024).

IOP Conf. Series: Earth and Environmental Science **342** (2019) 012003 doi:10.1088/1755-1315/342/1/012003

Research on Analysis and Application of Power System Stabilizer for Safe Operation of Large Nuclear Power Units

Huang Chao, Chen Weilong, Feng Weigang
Suzhou Nuclear Power Research Institute, Shenzhen Guangdong, 518124

huangchao2348@163.com

Abstract: As more and more large-scale nuclear power units(NPU) are put into operation, power system stabilizers(PSS) are important control means for units and power grids, which play an important role in the safe operation of NPU. In this paper, by analyzing the mechanism of accelerating PSS and the power regulation characteristics of NPU, the paper proposes a type selection of PSS for NPU. By analyzing the key technical requirements of the frequency link of the accelerating PSS, it proposes some test verification measures. Combined with the nuclear power plant reactor technical specification requirements, it studies the influence of the PSS on the nuclear island during the tuning test, and then proposes the technical conditions for the PSS setting test in NPU. In view of the large-scale use of semi-speed generators in NPU, it analyzes the effects between the torsional vibration frequency and PSS of semi-speed generators, and proposes new parameter settings and logic designs for PSS to solve similar problems.

1. Introduction

In order to ensure the safety of the power grid, the excitation regulators of the main power plants in the power grid have been put into the PSS in a large scale[1].

In addition to suppressing the low-frequency oscillation of the machine, the PSS also effectively suppress the regional low-frequency oscillation. The input of the PSS is a main measure to ensure the safety of the units and the power grid. Considering the problem of reversal of the prime mover [2], most of the PSS use the accelerating PSS in the units, called PSS2A/2B.

The safety of reactor power operation is the primary consideration for the safe operation of NPU. At the same time, NPU have the characteristics of large unit capacity and low steam parameters. At present, most NPU use semi-speed generator, so the use of nuclear PSS is special. In the design of the PSS and the on-site setting test of the NPU, in order to ensure the safe operation of the NPU, it is necessary to consider the particularity of the nuclear power unit.

This paper starts with the principle of accelerating PSS, then researches on the selection problem related to accelerating PSS and NPU operation characteristics, the key technologies for PSS frequency links related to safe operation of NPU, the nuclear reactor power limitation during PSS test, the relationship between torsional vibration and PSS setting of semi-speed generator, and finally proposes the corresponding solutions to ensure the safety of NPU.

2. Principle of the Accelerating PSS

The PSS2A model is mainly based on the evolution of the generator rotor dynamic equation. According to the mechanics law of the rotating object, the following relationship can be written for the rotor of the synchronous motor:

$$M\frac{d\omega}{dt}=T_m-T_e \quad (1)$$

Where: M is the moment of inertia of the unit rotor; Tm is the mechanical input torque; Teis the electromagnetic torque of the generator.

It is expressed by the per unit value and when the speed does not change much, the power is used to instead of the torque, then the above formula is changed to:

$$M\frac{d\omega}{dt}=P_m-P_e \quad (2)$$

After transposition of terms and given the deviation form, $d\omega/dt$is changed by S:

$$\Delta P_m=MS\frac{\Delta\omega}{\omega_0}+\Delta P_e$$

Because $\Delta\omega$ contains kinds of noise, both ends of the formula are multiplied by G(S) to filter the noise. The formula is expressed as:

$$\Delta P_m G(S)=(MS\frac{\Delta\omega}{\omega_0}+\Delta P_e)G(S)$$

The accelerating power is:

$$\Delta P_a=(M\frac{s\Delta\omega}{\omega_0}+\Delta P_e)G(S)-\Delta P_e \quad (3)$$

According to the formula (3), the transfer function of PSS2A is as follows:

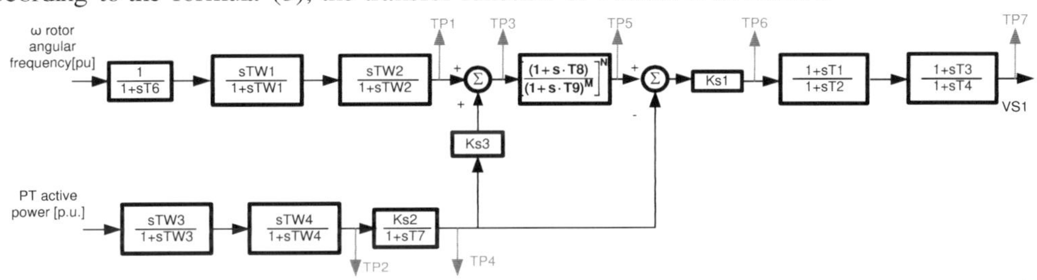

Figure 1. AcceleratingPSS(PSS2A)

ΔPa is the input of PSS, if the mechanical input power changed, the input signal of the PSS is the accelerating power ΔPa. If the mechanical input power never changed, then there would be such relation: ΔPm=0, ΔPa=ΔPe, it can be seen as the electric power input PSS.

3. PSS Selection of NPU

Using the accelerating PSS, when the mechanical power changes, the accelerating power output is approximately 0, so the PSS has no output, which eliminates the reverse modulation caused by mechanical power variation. When the small disturbance cause power oscillation in the system, the output of the prime mover remains unchanged, then the input signal Pm of the ramp function is 0, and the output signal should also be 0. Only the electric power single signal input works in PSS2A. At this time, the PSS2A and the single electric power signal PSS is consistent, and the PSS parameter compensation is consistent too. Through good phase compensation, it can suppress low frequency oscillations in the range of 0.2~2Hz.

Through the PSS2A model analysis, using the accelerating PSS can eliminate the reverse adjustment that occurs during the mechanical power adjustment process. At present, the second generation PWR nuclear power unit in China generally does not have the peak regulation. The possible maximum mechanical power change condition during operation is the RGL04 test or a circulating pump tripping, the load would be adjusted at a rate of 50 MW/min. Through the field test of the PSS2A and PSS1A models, the reverse adjustment of PSS1A is not obvious from the waveform of the RGL04 test. The comparison between the two type PSS are shown in Figure 2.

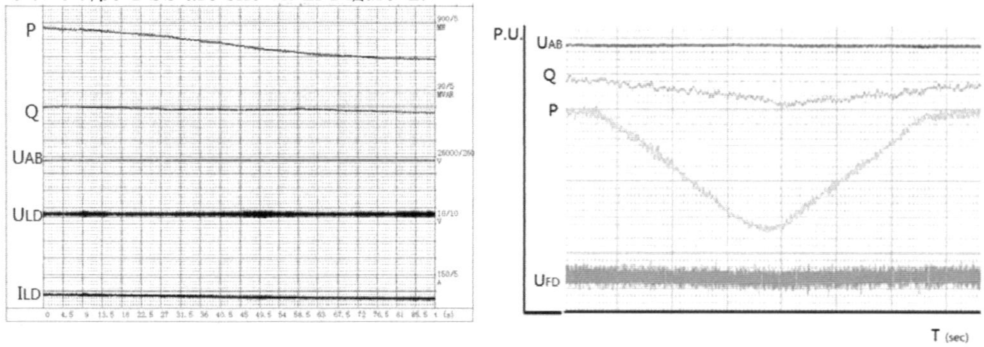

Figure 2. **(a)** Reverse test waveform of PSS2A **(b)** Reverse test waveform of PSS1

In Figure 2 (b), the unit is quickly reduced from 1000MW to 950MW at 50MW/min, the reactive power is changed from 27MVar to 18MVar; The unit is quickly recovered from 950MW to 1000MW at 50MW/min, and the reactive power is changed from 18MVar to 25MVar. During the whole test, the excitation voltage has no change, and the reactive power variation is less than 10MVar. There is no reverse adjustment in PSS during the normal operation of the unit.

From the above analysis, it is known that it is feasible to use the PSS2A model or the PSS1A model in NPU.

4. Analysis of Safe Operation of PSS in NPU

4.1. Frequency Link Analysis of the PSS2A

In order to eliminate the effects of mechanical power variations, a speed or frequency signal is introduced into the PSS2A model. In some excitation system supplied by an international supplier to some domestic NPU, the frequency link in the PSS2A model is not processed well[3].If there was a large interference, the phase of frequency link and the power link in PSS2A would not be matched well, which would cause negative damping and power oscillation.

The current microcomputer excitation system generally uses the generator line voltage and current to calculate the frequency of the internal potential of the generator. The principle is shown as follow: $E_q = U + jIX_q$.

From the engineering point of view, the change of the generator shaft speed is the same as the change of the internal potential frequency. By measuring the generator voltage U and the stator current I, the internal potential E_q of the generator can be obtained, and the generator shaft speed can be obtained by finding the frequency of the internal potential of the generator.

There are currently two methods to calculate the internal potential frequency of the generator[4],one is to calculate the frequency by means of the zero-crossing of the internal potential waveform and. The other is the angle of the internal potential, which requires the calculated phase current and line voltage to be 90 degrees. The defect of the second method is that the angle of the current and voltage would be disturbed by inrush current when the main transformer is charging. The above-mentioned abnormal PSS2A product uses this kind of frequency calculation method.

In the system oscillation, the mechanical power may have a very small magnitude change. Because of the large dynamic range, it must be handled correctly. If the excitation device was not handled well, frequency phase in PSS2A cannot well involved in model calculation, and an abnormality may occur. A lot of poor performance PSS is caused by insufficient hardware design or improper control parameter selection. Normally, it needs a CPU with rapid operation and high precision when designing PSS2A andit is better to use float algorithm when processing with algorithm. So, it is better to perform large external interference verification, and important small signals have sufficient accuracy verification in the field tests.

In addition, in order to verify the correctness of the frequency link, the two input signals and the output signal should have the following relationship in the field test: the integral phase of the frequency and the electric power is basically inverted, and the PSS output signal and the electromagnetic power signal are basically opposite (Figure 3) [5], these should be verified in the field test.

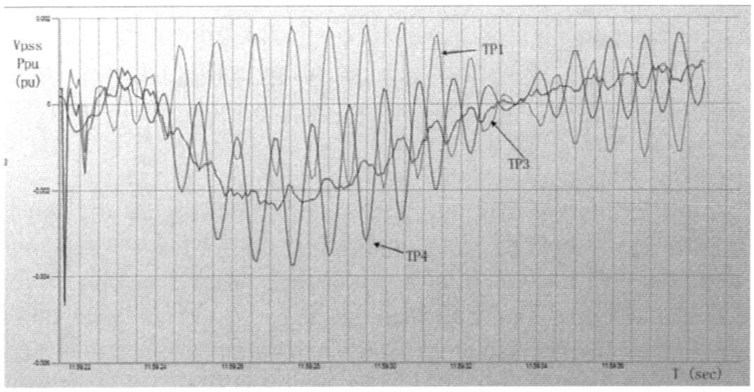

Figure 3. The phase relationship between the PSS output signal and power signal

4.2. The Impact on Nuclear Island During Step Test or Power Oscillation

In the normal operation of the generator, the output of the generator is coordinated with the output of the steam turbine. In the conventional thermal power plant, the output of the boiler is coordinated with the output of the turbine; In the NPU, the nuclear island is converted from the first stage pressure of the turbine to the thermal power constant to adjust the average temperature, the balance between the power of the nuclear reactor and the turbine output is regulated by the balance of the thermal power.

In the coordinated operation of the generator and the steam turbine, the electric power signal is sent to the steam turbine regulating system through the power transmitter to realize the coordinated operation between turbine power and electric power. According to the standard requirements of the power transmitter, the response time of the power transmitter is no more than 400ms. The response time of the latest microcomputer power transmitter for unit control is generally set as 50ms~90ms.When the generator excitation system performs a step test, the electric power fluctuation is within the response time of the power transmitter. In this process, the turbine power is not responding, which will not affect the nuclear reactor power. When the system experiences an increase amplitude or step test exceeded the response time of the power transmitter, the turbine power will respond. During the PSS step test, the generator power fluctuation peak time will cause a sudden change in the turbine power. Since the nuclear reactor power changes following the turbine output, when the following time was faster, it may cause nuclear island power changed. Therefore, the impact on nuclear reactor safety should be fully analyzed during the test.

Nuclear reactor technical specification requires: the nuclear reactor's rising power rate cannot exceed 3% PN/h in the first start after the nuclear power plant is assembled, when the full power stable operation time of the unit do not exceed 72 hours. When the PSS performs a 1%~4% step test under load, the fluctuation of electric power may exceed 3% PN/h. Therefore, it is necessary to avoid the time limit of the nuclear reactor's power rate limitation when conducting nuclear PSS tests.

4.3. Shaft Torsional Vibration and PSS Suppression Analysis of Semi-speed Generator

NPU's steam parameters are lower than conventional thermal power units. At home and abroad, some NPU use semi-speed generators, which use four pole-pairs. The inherent torsional vibration frequency of these mi-speed unit is generally lower than the full-speed unit. In the current operating nuclear power unit, some units have been found to have a torsional vibration frequency as low as 6 Hz, and some units have a torsional vibration frequency as low as 8 Hz. According to the IEEE 421 standard, it is necessary to consider that the inherent torsional vibration frequency of the unit may be as low as 4~5Hz[6].

The PSS adjusts the excitation voltage and changes the output reactive power of the generator by controlling the excitation regulator. In some cases, the PSS parameters cause the natural frequency of the shaft torsional vibration to be consistent or similar oscillating signal to be input to the rotor winding by the PSS, which may cause mutual excitation between the PSS and the shaft torsional vibration, resulting in sub-synchronous oscillation.PSS2A/2B is monotonically increasing in amplitude-frequency characteristics, with up-warping characteristics, which would cause the gain margin of the entire PSS link to be small and affects the gain value at the final steady-state operation. At the same time, in the high frequency or close to the natural frequency of the shaft torsional vibration, it has a boosting effect on the shaft torsional vibration. Therefore, considering the influence of the PSS on the shaft torsional vibration of the semi-speed unit, it would generally lower the high frequency gain value. There are two methods can be used: one is to filter out the signal above 3Hz in the second order low pass link of the PSS output, and needs to add extra links in the PSS software, and should also consider the influence of this link on the angle in low frequency; Another way is to use the PSS lead and lag compensation link, phase compensation should consider the whole frequency including the shaft torsional vibration frequency, and the phase of the low-frequency oscillation range and the shaft torsional vibration frequency should be compensated too.

For the lead and lag link $\frac{1+T5s}{1+T6s}$, when it is set as T6>T5, its phase-frequency characteristic is hysteresis and the amplitude-frequency characteristic is attenuated. According to this characteristic, for example [7], a compensation link with a corner frequency of about 3 Hz and a hysteresis angle of 15° is set, the frequency characteristics are as shown in Figure 4 , the gain is slowly attenuated to 76.7% of the original value at 3Hz, and after 3Hz the gain is rapidly attenuated, which can reduce the maximum gain to about 65% of the original value. [8] [9]

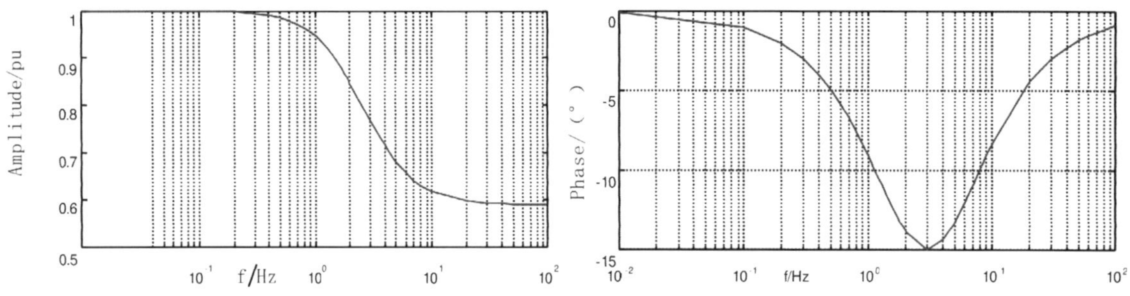

Figure 4. The bode diagram of first-order lag link

Considering the factors of the mechanical parameters which are affected by the shaft torsional vibration, the slope function in the PSS2A/2B model is used to reduce the PSS output deviation caused by the rapid change of mechanical power. At the same time, the ramp function also has the effect of attenuating the high frequency component of the input signal on the PSS output[10].Usually the high frequency component is above 7 Hz, and the torsional vibration frequency of the full-speed generator is above 10 Hz. After the ramp function is applied, the torsional vibration frequency of the full-speed generator is also filtered out. Since the steam parameters of NPU are lower than conventional thermal power unit, many nuclear power plants have adopted semi-speed generators at home and abroad. This semi-speed generator uses a 4-pole generator, and its torsional vibration frequency is as low as 4~5Hz. Therefore, when selecting T8 and T9 of the ramp functions, it should be noted that amplitude characteristic verification should be set according to the domestic typical value: T8=0.6s, T9=0.12s , M=5, N=1 (Figure 5).

From Figure 5 , it can be seen that in 4~5Hz, the amplitude of the ramp function is less than 0.1, correspondingly attenuating the signal of the mechanical combined power. Therefore, setting the ramp function according to typical values is also safe for nuclear power semi-speed generator.

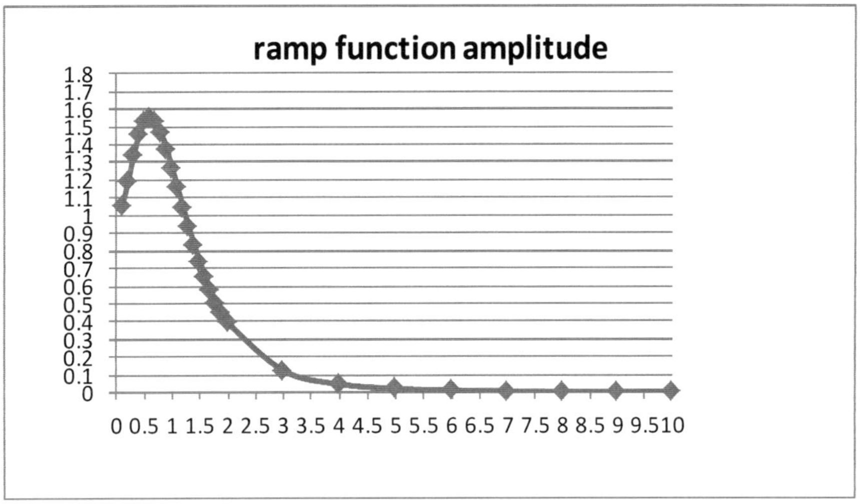

Figure 5. Ramp function amplitude frequency characteristics (x-axis as the frequency, y-axis as the ramp function gain)

5. Conclusion

This paper bases on the technical requirements of nuclear power plant reactors, nuclear generator characteristics and the use of PSS, and proposes the safety technical requirements for the use and test of PSS in NPU, then obtains the following conclusions:

1) there is no strict requirement in nuclear power units to use PSS2A model and PSS1A model;
2) In the PSS2A software and hardware design, it is necessary to verify the frequency link;
3) When carrying out the PSS step test, it is necessary to fully consider the technical requirements of the nuclear reactor technical specifications;
4) Power system stabilizer setting test needs to consider the natural frequency of torsional vibration of nuclear power half-speed unit, and take into account the influence of mechanical parameter operation on shaft torsional vibration. Through the supplemental or filtering of the ramp function parameters and the torsional frequency, the mutual excitation between the low torsional natural frequency of the half speed generator and power system stabilizer can be effectively reduced.

6. References

[1] Liu Qu. *Electric Power System Stability and Excitation Control of Generators[M].* China Power Press. 2007.03

[2] Zhu Shizhang, Chen Hao. *The Antiregulation of Turbine Power System Stabilizer[J].* Zhejiang Power. 2006, 5:9~11

[3] Wu Tao, Zhou Ji, Pu Jun,et al. *Analysis and test on the automatic voltage regulator of Three Gorges right bank power station [J].*Electric Power 2011,44(12)

[4] Xu Qipin, Hu Xianhong, Chen Xiaoming. *Methods for Realization of Dual-input Power System Stabilizer[J].* Hydropower Automation and Dam Monitoring 2008,32(5):8-11

[5] Wenyan Gu, P. Smulders, K. Mushens. *Power system stabilizer tuning-simulations and commissioning[C].* Power Engineering Society Summer Meeting.2000.IEEE.16 Jul 2000-20 Jul 2000

[6] IEEE Std 421.2-2014. *IEEE Guide for Identification, Testing,and Evaluation of the Dynamic Performance of Excitation Control Systems[S].* IEEE 2014

[7] Fang Sili,Su Weimin. *Disposition, Configuration,Parameter Calculation and commissioning of PSS[J].* Electric Power 2004,37(10):8-13

[8] Wu Jianchao, Huo Chengxiang, Dang Jie, etal. *Phase and Amplitude Coordination Optimization Method for Improving Gain of PowerSystemStabilize [J].* Automation of Electric Power Systems 2018,42(24)：203-208

[9] Zhang Fan, Xu Zheng. *Effect of Exciter and Power System Stabilizer on Subsynchronous Resonance Dampinf[J].*Transactions of China Electrotechnical Society 2007,22(6):121-127

[10] Mei Yong, Zhang Junfeng, Zhou Jian, et al. *Analysis and Improvement Measures on Power Oscillations in SilinHydropower Plant on July 6, 2017[J].* Automation of Electric Power Systems 2018,42(16):187-191

A Device for Online Monitoring of the Measurement Accuracy of Digital Electric Energy Meter

Yuchun Li[1], Dongdong Huang[1], Xudong Shi[1], Fajun Zhan[1], Xiaojian Liu[1] and Leijiao Ge[2]

[1] State Grid Xinjiang Electric Power Company Electric Power Research Institute, Urumqi, Xinjiang, 830011, China
[2] School of Electrical Information and Engineering, Tianjin University, Tianjin, 300072, China

Email: 50198766@qq.com

Abstract. The application of smart substation technology has brought many opportunities to the power industry, but also caused many new problems and challenges. At present, with the large-scale promotion of intelligent substation based on IEC61850, there are many engineering problems which will not appear in IEC61850 protocol and are closely related to practical application. A device for online detection of electrical error is designed in the paper for the measurement deviation problem of "IEC61850-9-2 input digital power meter" in the new generation of intelligent substation. The characteristic of the device is that once the meter measurement error is detected within the set time interval, it will record the voltage and current waveform data within the interval in real time for error analysis and problem finding.

1. Introduction

The accuracy of power metering is directly related to the economic interests among power generation enterprises, power grid companies and electricity customers. All metering devices must be checked before they are connected to the power grid. Even if it is verified and qualified in the laboratory, there will still be a large deviation after being connected to the network because the research of electrical energy metering based on IEC61850 is not mature enough presently in [1-6]. The reasons for this situation may be manifold, possible reasons include but are not limited to the following:

- The synchronization of data is not well handled when merging unit works in [7].
- Some of IEC61850 9-2 message are lost or jitter in the transmission process.
- The periodicity problem of power calculation is not properly handled when frequency deviates.
- The influence of harmonics on electric energy metering.

The traditional analog metering system is composed of electromagnetic transformer and electric energy meter with analog input, connected by cable. Its errors mainly come from analog-to-digital conversion and algorithm on the electric energy meter side in [8]. In the smart grid, the analog-to-digital conversion is completed at the transformer or merging unit, and the sampled digital quantity is transmitted to electric energy meter through fiber-optic. The introduction of new error characteristics makes traditional methods no longer applicable. Thus, the paper designs a device which can monitor the error change of digital power meters in real time in a set time interval. When the detected error is beyond limiting value, the original signal waveform and 9-2 message in this time interval are saved.

2. Design of power monitoring device system

In the traditional analog metering system, it is assumed that there is no delay in the transmission process, so the voltage signal and current signal are naturally synchronized when they are connected to electric energy meters. However, in IEC61850 digital station, due to the delay of different sampling devices and the resampling in the merging process, the synchronization of digital voltage and current sampling values becomes an important problem [9]. This problem exists in the two classical structures of digital substation metering system at present. In the application of electronic transformer and merging unit, the sampling unit is embedded in the transformer, so the voltage sampler and the current sampler are isolated. They eventually enter the electric energy meter through the same SMV message which will inevitably pass through the merging unit. The other one is that in the application of the traditional electromagnetic transformer and analog sampling value merging unit, the traditional transformer secondary signal is directly connected to the merging unit, that is, the input of the input merging unit is "numerically synchronized", then its outputs are synchronized.

When the digital sampling values of input in electric meter are not synchronized, the measurement of the digital power meter will be abnormal inevitably. For this problem, it is necessary to assume that the input of the merging unit is unreliable when considering the accuracy of online measurement of electric energy meter. So the device designs an electric energy calculation unit for the analog signals before the merging unit, and takes the calculated electric energy as the benchmark to determine whether the electric energy meter value is out of tolerance or not.

Structurally speaking, the digital electric energy meter with IEC61850 9-2 message does not have sampling unit itself. Its errors mainly come from algorithm error and interference of communication fault factors. In order to better record the relationship between communication faults and point energy anomalies, another main function of the device is to record the digital message according to the measurement comparison results. The device includes three power metering modules and a recording module. The structure diagram of the device is shown in figure 1.

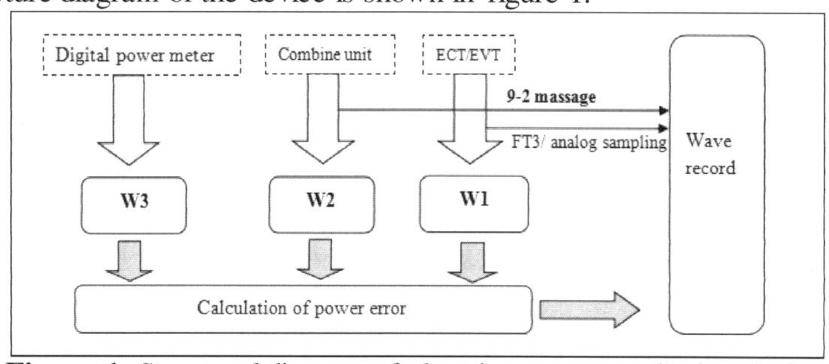

Figure 1. Structural diagram of electric energy monitoring device.

The working process of the device is as follows: Firstly, according to the output of the transformer, calculate the electric energy W1 without merging unit merging. Then, calculate the electric energy W2 based on the SMV message output by the merging unit. Finally, get the measurement result W3 based on the pulse signal output by the digital power meter. When there is a big deviation among three electric values, the recording unit is notified to retain the input of each part.

3. Design of electric energy metering and verification hardware

STM32 and FPGA are used in the bottom layer of the device. Exynos4412 is used to drive SATA hard disk to complete waveform recording and drive 10.1-inch capacitive screen to realize human-computer interaction. The three parts of energy metering unit use common STM32 and FPGA hardware resources. Electric energy metering mainly counts the output pulse of the electric energy meter. Pulse output modes can be divided into two categories: active output and passive output in [10]. The active output mode does not require the terminal to provide power, the pulse voltage source range is 5±5V, the pulse width is 80±20ms, the meter and the 0.20 terminal have electrical common points. The passive output mode requires power supply from the terminal. The meter and the terminal are separated by optical coupling. The pulse voltage amplitude is same as the power supply provided by terminal, the pulse width is 80± 20ms.

3.1. Calculation of the output electric energy of the transformer

The merging unit of analogical input is still widely used, so the main consideration is to calculate electric energy from analogical signal. The hardware block diagram for calculating the analogical power is shown in figure 2.

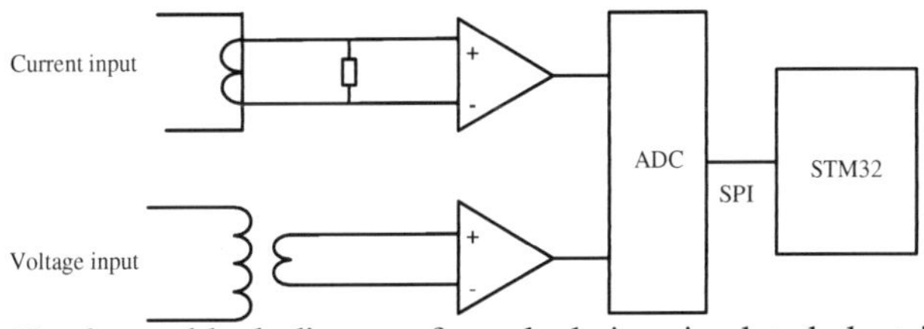

Figure 2. Hardware block diagram for calculating simulated electric energy.

The input voltage signal is transformed into a 2.5V signal by the transformer inside the device and input into the PGA amplifier unit. When the converted signal is less than the ADC resolution, STM32 adjusts PGA to enlarge the signal and then sampling it. The voltage sampling process is similar. Through sampling values calculated by suitable algorithms, the electric energy carried by the analogical semaphore output by the transformer can be obtained.

3.2. Calculation of merging unit output electrical energy

According to IEC61850 protocol, when the sampling rate is 4000 samples/ sec, the interval between sampled data messages is 250us. The protocol stipulates that the TLV encoding format is used when the sampled value message is transmitted by Ethernet. The TLV format is shown in figure 3.

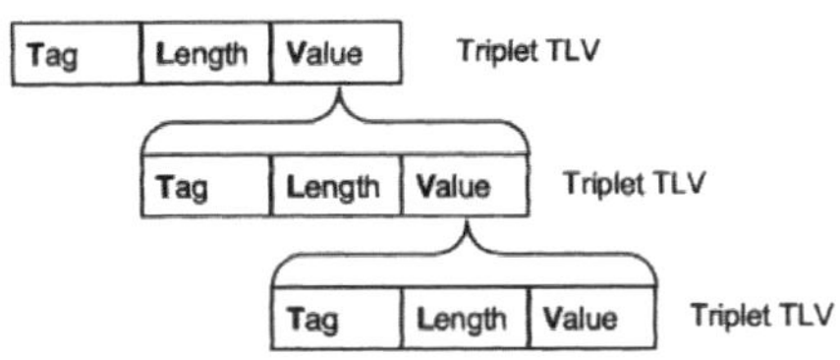

Figure 3. TLV encoding format.

The message using the TLV encoding structure is nested layer by layer. The lower layer exists as the upper layer of value, that is, the upper layer of value may exist paralleled with one or more TLV blocks. At the same time, each TLV block has its own tag, length and value. Therefore, when using this structure, the location of any value block is not fixed in different applications. The requirement of hardware processing speed and real-time performance for receiving 250us Ethernet messages is very high. Finally, the device chooses the scheme of directly driving PHY chip by FPGA as the hardware of acquiring W2.

4. Software design of wave recording unit

The system consists of four modules. Message parsing module receives 9-2 signals from merging unit and extracts sampled data from message. Analog sampling is completed by W1 acquisition unit, the sampled data will be sent to the value recording unit through the instrument's internal communication line. Data synchronization and recording module stores the parsed digital sampling value data and the

IOP Conf. Series: Earth and Environmental Science **342** (2019) 012004 doi:10.1088/1755-1315/342/1/012004

analogical sampling data sent on the bottom layer synchronously. Data management unit mainly manages the stored waveform data. The block diagram of this part is shown in figure 4.

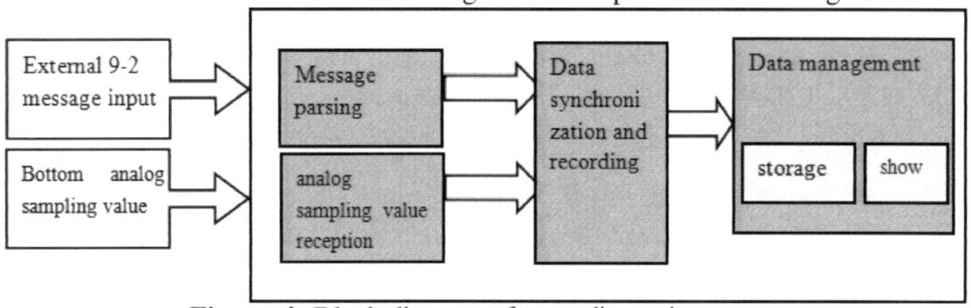

Figure 4. Block diagram of recording unit system.

The traditional analog recording devices only need to sample and store analogical signals on copper cables, while the IEC 61850 recording device also needs to have the analytical ability of sampling value message and data model. According to the protocol, SV messages are packaged in ISO/IEC 8802-3 frame format and then transmitted in 100-basefx network, which is shown in the figure 5.

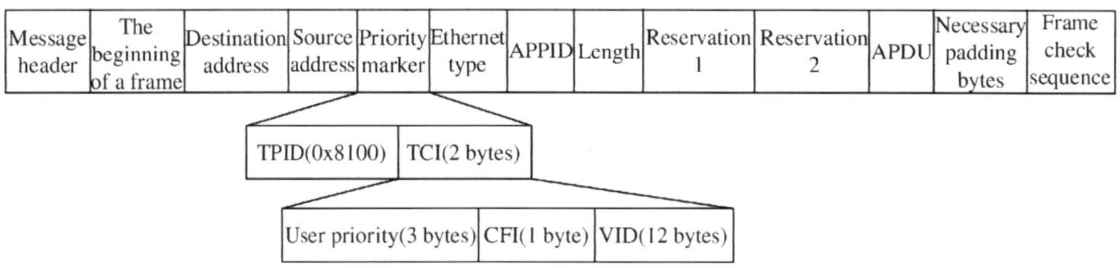

Figure 5. ISO/IEC8802-3 message data frame format.

The type of this message is defined as a broadcast message whose broadcast address is defined as 01-0C-CD-04-00-00 and 01-0C-CD-01-01-FF. VLAN part is added on demand, the ethertype is 0x88BA, APPID is selected between 0x4000-0x7FFF. Length represents the size of the entire Ethernet packet data, and its value is equal to the actual number of bytes occupied by five parts of APPID, length, Reservation1, Reservation2 and APDU. Reservation 1 and reservation 2 are both 00H. The actual distribution of APDU is shown in figure 6.

APDU is used as the actual data part of Ethernet message after passing ASN.1 /BER, it contains only one application service data unit. SmpCnt is a sampling value serial number, ConfRev is the configured version number, SmpSynch is a synchronization marker, svID value type is a string. The Data is the voltage and current sampling value. 8 sets of sampling values in the figure are the value channel, from top to bottom are respectively channel 1 to channel 8.

Since the input of waveform data comes before the judgment result, cyclic recording is adopted in this paper. The specific steps are as follows:

- When the comparison starts, the timer is started. When the timer ends, inform the power comparison and the recording part respectively and stop the action.
- According to the set message parameters, the 9-2 analysis module analyses SmpCnt and sampled voltage and current data needed. The parsed data is stored as a temporary file. The analogical sample value data is transmitted to ARM processor by internal communication mechanism. After analysis, the data is processed in the same way as 9-2 data.
- 9-2 data and analogical sample value data are stored by the same storage module in chronological order for comparison in use.
- Power comparison module performs electric energy comparison and timing according to W1, W2, and W3, and notifies the recording module when the predetermined time is reached.

IOP Conf. Series: Earth and Environmental Science **342** (2019) 012004 doi:10.1088/1755-1315/342/1/012004

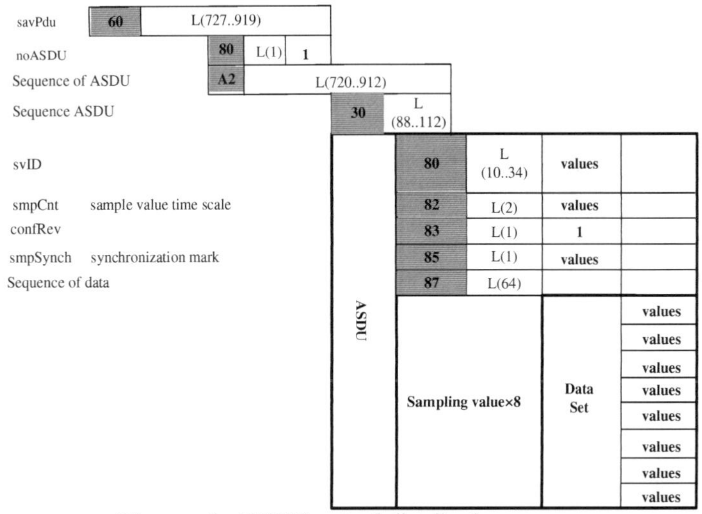

Figure 6. APDU actual distribution map.

- After reaching the predetermined time, the recording module decides whether to permanently store waveform data based on whether the electrical energy comparison is abnormal.

5. Summary and prospect

The application of this device in the digital metering simulation laboratory does provide a good reference for finding out the fault of metering, and greatly facilitates the positioning of the causes of energy metering errors in the digital substation metering system. However, there are still some limitations in the use of the device. In the future, the improvement of measuring accuracy of equipment in various practical application scenarios can be studied in depth.

6. References

[1] Sun W M and Lin G Y 2010 *J.* Research on calibration method of IEC61850 protocol watt-hour meter *Electrical measurement and instrument* **47** 13-15

[2] Li B and Ma C 2010 *J.* Distributed bus protection scheme based on IEC61850 *Power System Automation* **34** 66-69

[3] Li J G and Wang Y Y 2010 *J.* Research and application of IEC61850 standard in fault recording device *Power system protection and control* **38** 97-99

[4] Sun H, Yuan H M and Lu S M 2009 *Int. Conf. on Information technology and computer science* vol 1 (Kiev: Ukraine) pp 264-267

[5] Pan F, Lin G Y, Wang L Y and Song Q 2017 *13th IEEE Int. Conf. on Electronic Measurement & Instruments* (China: Yangzhou) pp 466-472

[6] Hou X Z, Chen W L, Tao X D and Hu J 2018 *13th IEEE Int. Conf. on Electronic Measurement & Instruments (ICEMI)* (China: Yangzhou) pp 139-144

[7] Zhu J B and Kong H L 2016 *J.* Discussion on merging unit applied to ower quality monitoring in intelligent substation *Computer science and exploration* **10** 262-267

[8] Sheng Q G 2012 *J.* Error analysis and test method optimization for constant test of watt-hour meter *Electrical measurement and instrument* **49** 55-59

[9] Tome MD, Nardelli PHJ and Alves H 2019 J. Long-range low-power wireless networks and sampling strategies in electricity metering *IEEE trans on industrial electronics 66-2 pp1629-37*

[10] Zheng P X and Lai G S 2005 *J.* Discussion on monnection mode between load management terminal and pulse output interface of electric energy meter *Fujian electric power and electrotechnical* **25** 40-41

A New Type of Hybrid DC Current Limiting Circuit Breaker suitable for HVDC

J J Wang[1], Z X Wang[1*]

[1] Electronic Information and Electrical Engineering Department, Shanghai Jiao Tong University, China

*Corresponding author: wangzxin@sjtu.edu.cn

Abstract. Multi-terminal DC transmission technology based on voltage source converter is one of the important technical means to solve the problem of grid connection and consumption of renewable energy. In this regard, encountering the fault current during fault period remains an important issue. Thereby, devices and circuits to suppress the fault current have become a hot topic in recent years. In this paper, a new type of current-limiting circuit breaker structure is proposed. Under normal conditions, the load current flows through the low-loss main branch. After the fault occurs, the power electronics based switches are disconnected, and the fault current is shifted to the auxiliary branch of current limiting device. The current limiting device suppresses the rise of the fault current. Finally, the power electronics based switches in the auxiliary branch open, and eventually the residual current is absorbed by the MOA. Moreover, the proposed topology of current-limiting circuit breaker is verified using PSCAD/EMTDC based simulation experiments. In comparison with ABB circuit breaker, the proposed topology has small on-state losses and good damping characteristics under normal conditions, and in the fault state too, Besides, the proposed topology owns the benefit of high reactance, therefore, allow the suppression of fault current effectively and have advantage to cut-off large amount of fault current.

1. Introduction

Large-scale applications of fully controlled power electronic devices result in the DC transmission based on voltage source converters (VSC), which mitigate the shortcomings of conventional high voltage DC transmission based on Line-Commutated Converters (LCC). The utility of VSC-HVDC allows the capitalization of renewable energy resource and smart grid links more effectively. One of the biggest advantages of VSC-HVDC is the independent control of both active and reactive power [1-4]. However, in the cause of fault at any point in the DC transmission line, the DC link capacitors discharge rapidly. As a result, DC current increases exponentially and system voltage decay sharply, these implications can cause huge damage to power converters at the converter station [5-8]. It is the obvious reason, that protection of DC grid is still the hot topic for research.

The means of handling the DC side short-circuit fault can be divided into three types according to the processing area [9-11]: 1) The AC circuit breaker which is used to cut off the connection between the DC network and the AC system on the AC side, 2) The self-clearing of the DC side short circuit fault realized by the converter itself control and 3) The DC circuit breaker in DC side. The first two methods can effectively suppress the fault current fed by the AC side, but cannot suppress the inrush current generated by the DC link capacitor discharge. The DC circuit breaker can effectively suppress the inrush current of the capacitor discharge, but there are some technical difficulties in the development of the high-voltage DC circuit breaker, mainly the high breaking capacity and the short-time cutting fault.

At the end, Reference [12] proposed a new type of circuit breaker topology with fault current limiting function. The circuit breaker is composed of three branches with inductance and power electronic switching devices. The topology has functions such as open circuit and current limiting, and has advantages that on-state loss is small, the short-circuit current is large, and the voltage level is high. Reference [13] proposed a rectified hybrid high-voltage DC circuit breaker topology, which uses a bridge-type reversing valve group and a one-way switch valve group to form a branching branch. This topology uses half the number of insulated gate bipolar transistors compared with the conventional topology, and can realize bidirectional current breaking, which is economical and reliable. Reference [14] proposed a circuit breaker topology with forced commutation circuit, which can block the fault current within a few milliseconds, and the topology realizes the zero current open circuit of the mechanical switch. The circuit breaker can maintain the conduction state even if the semiconductor devices suddenly experiences blackout.

This paper proposes a new type of current-limiting high-voltage circuit breaker topology. Then analyses the stress characteristics of the fault current, and research the influence of the selection of resistance and inductance parameters of the current limiting device on the fault current. Finally, the rationality of the proposed method is verified by the simulation model in PSCAD/EMTDC.

2. Working principle

2.1. Topolopy

The topology of the circuit breaker proposed in this paper is shown in Figure 1. Among the components of proposed topology, L1 is a current limiting inductor. Whereas, the auxiliary branch is composed of single-conducting IGBT tubes. The bidirectional shutdown of the current is realized by the rectifier bridge. The auxiliary branch can also be directly guided by a pair of IGBT tubes, as shown in Figure 2. The auxiliary branch is connected in parallel with the MOA, and the MOA protects the IGBT tube from overvoltage. A current limiting circuit is connected in series with the auxiliary branch. It is important to mention that, the current limiting circuit is composed of an inductor and a resistor in parallel as depicted in Figure 1 and Figure 2. When the fault current is transferred from the main branch to the auxiliary branch, the current limiting circuit suppresses the rise of the current and seeks more time for cutting off the fault.

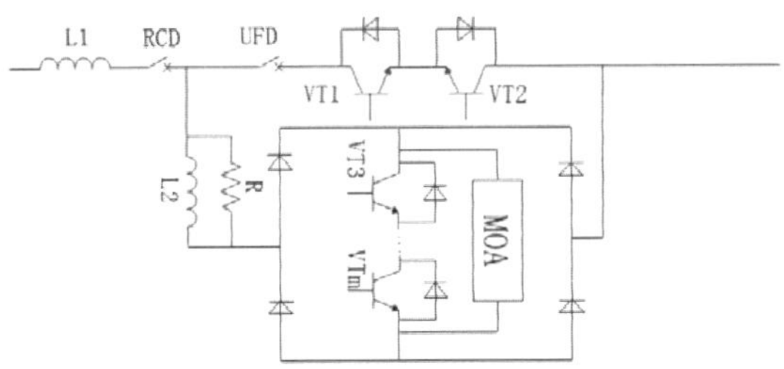

Figure 1. Rectifier-based circuit breaker topology

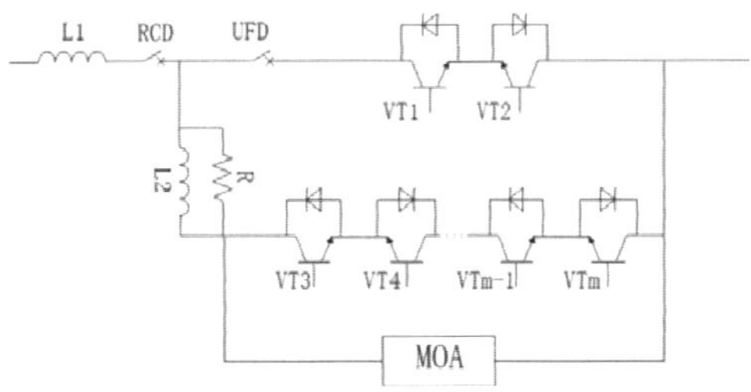

Figure 2. Circuit breaker topology using paired IGBTs

2.2. Working principle

The operation of the new current limiting high voltage circuit breaker can be divided into three phases. The first phase is the fault detection phase, and the fault current flows through the main branch. The second phase is the current transfer phase. The power electronic device switch of the main branch is disconnected, the IGBT of the auxiliary branch turned on, then the fault current is transferred from the main branch to the auxiliary branch, and the ultrafast disconnector of the main branch is disconnected. On the third, the switches of stage auxiliary branch are opened, the current is absorbed by the MOA, and the current will be reduced to 0, then the fault can be removed.

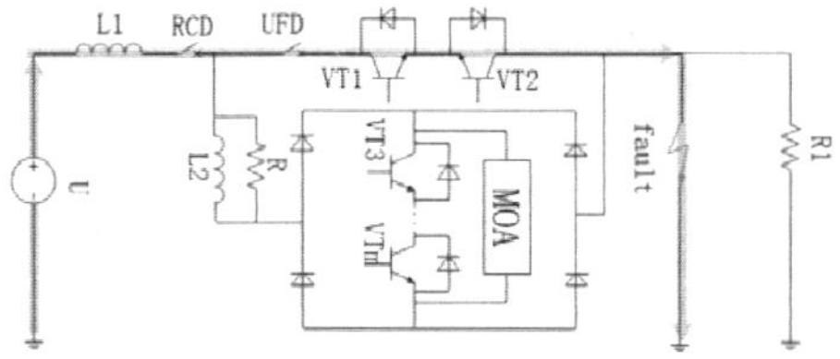

Figure 3. Current flow during t_0-t_1 period

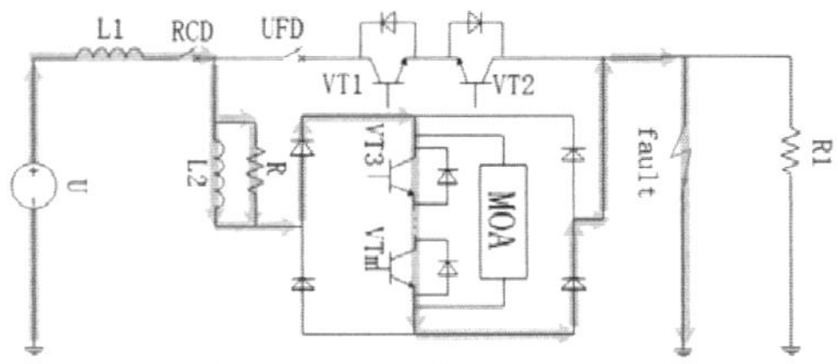

Figure 4. Current flow during t_1-t_2 period

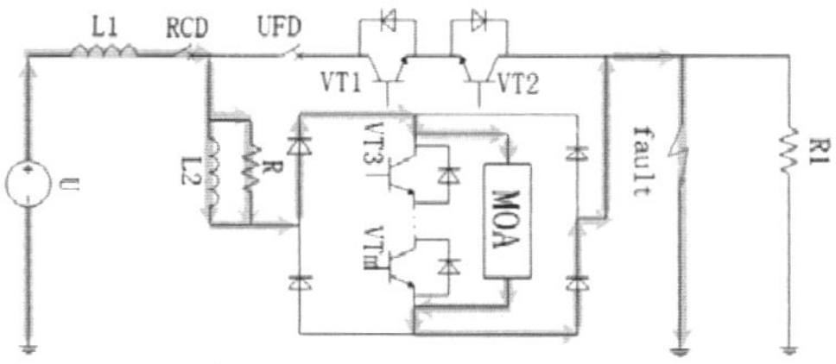

Figure 5. Current flow after t_2

2.2.1. System Operation During t0≤t<t1. The flow of current during $t_0 \le t < t_1$ is elucidated in Figure 3. Ignore the conduction loss of the IGBT and the ultrafast disconnector. The circuit breaker can be equivalent to an inductor. The current of the line is as follows:

$$I = \frac{U}{L_1}\left(t - t_0\right) + \frac{U}{R_1} \tag{1}$$

The ABB circuit breaker has the capacity to the cut-off current of 9kA and rate of fault current raise is less than 5kA/ms. The parameters of the circuit breaker in this paper are also designed according to this, and two constraints can be obtained:

$$\begin{cases} I(t_1) < 9\text{kA} \\ {}^{\mathrm{d}I_1}\!/_{\mathrm{d}t} < 5\text{kA/ms} \end{cases} \tag{2}$$

According to Equation 2, the range of values of the inductor L1 can be obtained.

2.2.2. System Operation During t1≤t<t2. The flow of current during $t_1 \le t < t_2$ is highlighted in Figure 4. Neglecting the on-state losses, power electronics based switches have no impact on fault current. Application of KVL yields in equation (3):

$$L_1 sI\left(s\right) + \frac{R_2 L_2 s}{R_2 + L_2 s} I\left(s\right) = \frac{U}{s} + L_1 i\left(t_{1^-}\right) \tag{3}$$

Assuming that the initial current of the inductor at time t1 is k, the fault current is available:

$$I\left(t\right) = \frac{U}{L_1 + L_2} t + \frac{R_2 L_1^2 k + R_2 k L_1 L_2 + U L_2^2}{R_2\left(L_1 + L_2\right)^2} - \frac{R_2 L_1^2 k + R L_1 L_2 k + U L_2^2}{R_2\left(L_1 + L_2\right)} e^{\frac{L_1 R_2 + L_2 R}{L_2 L_1} t} \tag{4}$$

In equation (4), $I(t_1^+)=I(t_1^-)$ can be found, so the current is continuous and the inductor does not generate an overvoltage. Deriving the current leads to equation (5):

$$I\left(t\right)' = \frac{U}{L_1 + L_2} - Ce^{\frac{L_1 R_2 + L_2 R_2}{L_1 L_2} t} \tag{5}$$

Where C is $\dfrac{R_2 L_1^2 k + R_2 L_1 L_2 k + U L_2^2}{L_1 L_2\left(L_1 + L_2\right)}$, from Equation 6, it can be found that the current is first decreased and then increased. Therefore, in the design of the inductor and resistor parameters, the minimum value of the current should be as close as possible between t_1 and t_2.

2.2.3. System Operation for t≥t2. After the auxiliary branch switch is turned off, the MOA absorbs the residual current and the current drops rapidly to zero. The current flow path at this stage is shown in Figure 5.

3. Simulation and results

To verify the theoretical implications of the proposed topology of the circuit breaker, a series of simulation-based experiments were conducted using PSCAD/EMTDC. Table 1 represents the values of different components used for simulation purpose.

Table 1. Parameters of each component.

U_{dc}/kV	R_1/Ω	R_2/Ω	L_1/H	L_2/H
500	166.7	100	0.17	0.4

Assume that a fault at 0.2 sec occurred in the system, power electronic switches in the main branch are turned off at 0.202 sec. Likewise, the power electronic switches in the auxiliary branch are turned off at 0.205 sec. In order to study the influence of resistance in the current limiting circuit on fault current, the value of resistances was changed to 10, 50, 100, 1000 ohms respectively, while other parameters of the system remained unchanged. The line current under influence of different resistive values is depicted in Figure 6.

In Figure 6, it can be observed that larger the resistance smaller the value of maximum current and quicker the decay time. However, if the resistance is too large, the inductor can cause an overvoltage. It must be noted that the resistance cannot be chosen too small, because when the resistance is less than a certain value, the current transferred to the auxiliary branch will not drop and will continue to rise. The value of the resistor should be as close as possible to the point where the current reaches the minimum value point in the middle of the auxiliary branch conduction period.

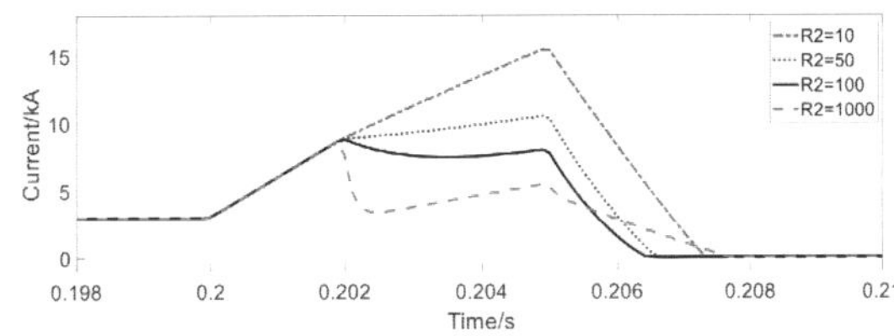

Figure 6. The curve of current under different resistance

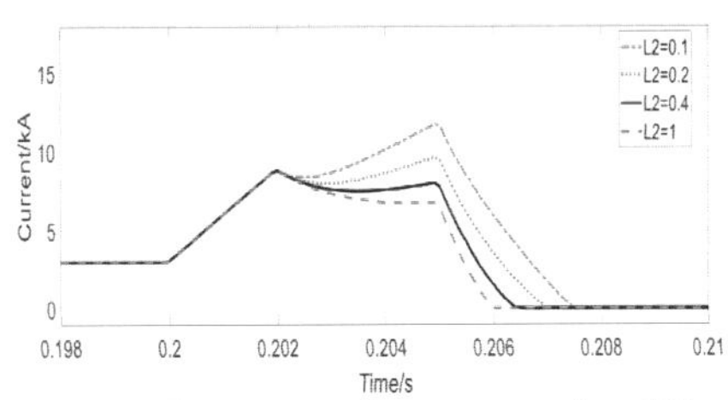

Figure 7. The curve of current under different inductance

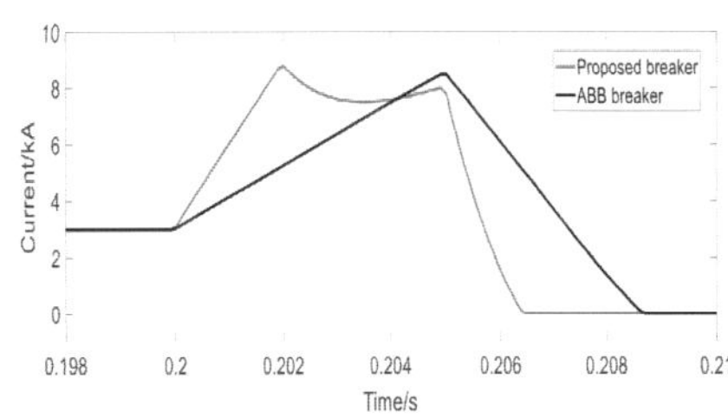

Figure 8. Simulation result comparison chart

Similarly, in order to study the influence of the inductance change (on the fault current) in the auxiliary branch of the current limiting circuit, the value of inductance is changed, the values chosen for inductance are 0.1, 0.2, 0.4, 1H respectively, whereas, other parameters remained unchanged. The line current under influence of different inductances is depicted in Figure 7.

It can be seen from Figure 7 that the inductance mainly affects the rate of change of current. The ABB hybrid circuit breaker (inductance value 0.45H) is compared with the scheme proposed in this paper. The simulation results are shown in Figure 8.

It can be found that although the circuit proposed in this paper has a faster current rise between 0.2 and 0.202 s, the current suppression effect is much better than that of the ABB circuit breaker between

0.202 and 0.205 s. At the same time, the energy dissipation is faster than ABB breakers. Moreover, under normal operating conditions, the scheme of this paper has less on-state loss and better dynamic characteristics.

4. Conclusion

A new topology of the circuit breaker with current limiting capability was purposed in this paper. Theoretical analysis inclusive of working principle, parameters estimations and influence of different components was backed by simulation based experiments on PSCAD/EMTDC, which mimic the correctness of proposed topology. The topology discussed in this paper owns the following distinct advantages: 1) In normal operating condition the inductance of the line is relatively small, as a result loss is small too, and the dynamic characteristics are good. 2) Under the fault condition, the maximum value of current is not too high, besides, the rate of rise of current during the fault is also minimized effectively.

The future work will be to study the calculation method of the key parameters of the circuit breaker and verify it in the multi-terminal flexible DC transmission model.

5. Reference

[1] Xu Z and Chen H 2007 Review and applications of VSC HVDC [J] *High Voltage Engineering* **1** 000
[2] Flourentzou N, Agelidis V G and Demetriades G D 2009 VSC-based HVDC power transmission systems: An overview *IEEE Transactions on power electronics* **24** 592-602
[3] Bahrman M P and Johnson B K 2007 The ABCs of HVDC transmission technologies *IEEE power and energy magazine* **5** 32-44
[4] Bresesti P, Kling W L, Hendriks R L and Vailati R 2007 HVDC connection of offshore wind farms to the transmission system *IEEE Transactions on energy conversion* **22** 37-43
[5] Yang J, Fletcher J E and O'Reilly J 2012 Short-circuit and ground fault analyses and location in VSC-based DC network cables *IEEE transactions on Industrial Electronics* **59** 3827-37
[6] Rafferty J, Xu L and Morrow D J 2012 DC fault analysis of VSC based multi-terminal HVDC systems
[7] Li R, Adam G P, Holliday D, Fletcher J E and Williams B W 2015 Hybrid cascaded modular multilevel converter with DC fault ride-through capability for the HVDC transmission system *IEEE Transactions on Power Delivery* **30** 1853-62
[8] Bucher M K and Franck C M 2013 Contribution of fault current sources in multiterminal HVDC cable networks *IEEE Trans. Power Del.* **28** 1796-803
[9] Candelaria J and Park J-D 2011 VSC-HVDC system protection: A review of current methods. In: *Power Systems Conference and Exposition (PSCE), 2011 IEEE/PES*: IEEE) pp 1-7
[10] Elserougi A A, Abdel-Khalik A S, Massoud A M and Ahmed S 2014 A new protection scheme for HVDC converters against DC-side faults with current suppression capability *IEEE transactions on power delivery* **29** 1569-77
[11] Franck C M 2011 HVDC circuit breakers: A review identifying future research needs *IEEE Transactions on Power Delivery* **26** 998-1007
[12] Li S, Zhang J, Xu J and Zhao C 2019 A new topology for current limiting HVDC circuit breaker *International Journal of Electrical Power & Energy Systems* **104** 933-42
[13] Shi Wei, Cao Dongming, YANG Bing, Lv Wei, Wang Wenjie and Liu Bin 2018 500 kV Commutation-based Hybrid HVDC Circuit Breaker *Automation of Electric Power Systems* **42** 102-107.
[14] Bingjian Y, Yang G, Xiaoguang W, Zhiyuan H, Longlong C and Yunhai S 2015 A hybrid circuit breaker for DC-application. In: *DC Microgrids (ICDCM), 2015 IEEE First International Conference on*: IEEE) pp 187-92

Acknowledgments

This work was supported by National Key R&D Program of China 2018YFB (0904600), Shanghai Science & Technology Commission Program(18DZ2205700); Science & Technology Program of Minhang District of Shanghai of China (2017MH271).

IOP Conf. Series: Earth and Environmental Science **342** (2019) 012006 doi:10.1088/1755-1315/342/1/012006

Study on PV Panel Cooling System using IoT with ESS for Preventing Reduced Efficiency of Solar Panel

Inho CHO, Heeje KIM*

School of Electrical Engineering, Pusan National University, Busandaehak-ro 63beon-gil, Geumjeong-gu, Busan, 46241, South Korea

*Corresponding Author E-mail: heeje@pusan.ac.kr

Abstract Solar photovoltaic systems are renewable energy sources that are used around the world. However, one of the causes of reducing the efficiency of solar systems is the temperature increasing in solar panels. To prevent this phenomenon, a cooling fan can be installed on the back side of the solar panel. The solar system efficiency also decreases due to weather conditions and unexpected situations. To overcome this problem, an IoT (Internet of Things) system was used to monitor the state of the solar system and control the cooling fan. The core microprocessor used in IoT systems was Arduino. Using Arduino, an IoT system can be implemented simply and economically. Also by following the paradigm of using ESS(Energy Storage System) smartly, we can use long-lived supercapacitor to operate the fans. The entire system was designed and tested and the efficiency increased by approximately 4.7%. Although it is a small 30W capacity photovoltaic system, its efficiency is expected to be increased by applying it to a photovoltaic system of more than 1kW in the near future.

1. Introduction

Encapsulation alters the heat flow into and out of the PV module, thereby increasing the operating temperature of the PV module. These increases in temperature have a major impact on the PV module by reducing its voltage, thereby lowering the output power. Recent studies have shown that the efficiency of an amorphous silicon cell decreases by 0.05% for every 1 °C increase in temperature and in the case of a crystalline silicon solar cell, the decrease ranges from 0.4% to 0.5% [1]. As shown in Figure 1, the output power decreases with increasing temperature.

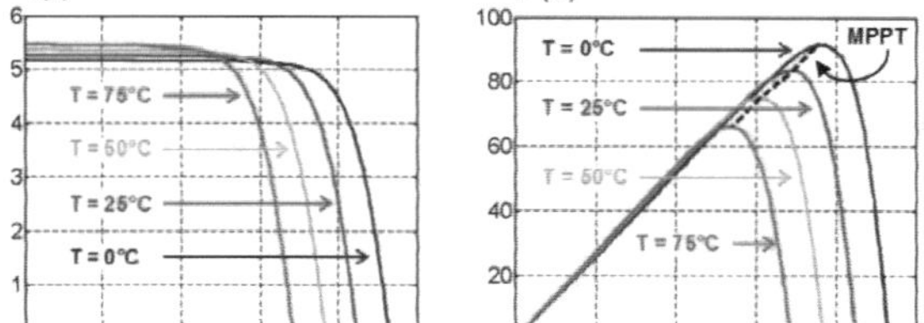

Figure 1. Solar panel output power with temperature change

Although the temperature differs according to the season and location, the temperature of a solar panel is usually highest in summer [2]. Various solutions have been proposed to prevent the efficiency decrease due to such a temperature rise. There are various method to deal with efficiency problem, in this study, we choose the cooling fan method. The cooling fan is turned on and off automatically by the temperature of the solar panel. On the other hand, operating the cooling fan in the event of clouds or inverter failure is inefficient. Therefore, in some situations, real-time remote control is needed.

And if we can apply ESS in the setup(Even we haven't yet) to operate fans, we can check the battery level in real time, in some case with needs, use electric energy from panel directly to charge it. It will help to simplify the energy managing algorithm. And If we use supercapacitor for it, we will see the positivity even not to care about life time of all device setup cause of batterys' long life time.

To implement such a system, an operating system based on the IoT (Internet of Things) is required. The IoT is an ecosystem of connected physical objects that are accessible through the internet. This is next generation technology that will soon be applied to society. [3] The IoT refers to the technology or environment, in which sensors are attached to transmit data to the Internet in real time. [4-5] Based on the IoT, the temperature of the solar panel can be controlled anytime and anywhere through a smart phone. In this paper, the ultimate goal is to automatically control the temperature of the solar panel using an IoT system.

2. Devices and Methods

The devices used in this experiment were divided into three parts. First, it receives the temperature data through a temperature sensor as the main control part, activates the relay through an external interrupt, and controls the overall system operation. The second is the solar panel part. This part receives the signal from the relay and activates the cooling fan installed on the back of the solar panel. The final is a smart phone application part that allows the users to monitor solar power system states and control the cooling fan whenever and wherever. The parts and mechanisms of each of these three parts have been described.

2.1. Control part

This part receives the various data values as the main control part and transmits the data to the server in connection with the internet network. This also activates the cooling fan on the back of the solar panel through an external interrupt. The microcontroller used was the Arduino Mega2560. This device was optimized to implement the IoT and is economical and can be interworked with a range of sensors. A WIFI shield (JSN270) for implementing the IoT was also used.

The power to operate the Arduino was obtained from an internal power source produced from a solar panel. The power consumed by the control part was 1.12 W. Figure 2 shows the internal view of the control box. In the center, there are the Arduino and WIFI shield as well as converters, current sensors, and relays around.

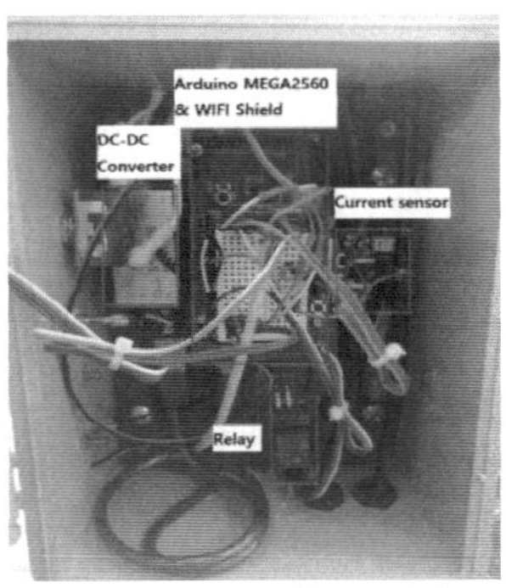

Figure 2. Inside control box with Arduino and WIFI shield

Arduino controls the system via the flow diagram below. The cooling fan turns on automatically when the temperature of the solar panel exceeds 40 °C and when there is no cloud. Figure 3 presents a flow diagram of this system.

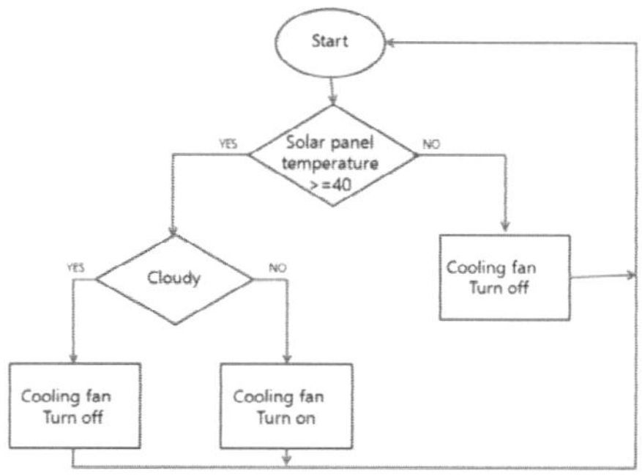

Figure 3. Flow diagram of this system

2.2. Solar panel part

This part consists of a cooling fan and a solar panel. Four cooling fans are installed on the back of the solar panel. This is arranged in a regular Square arrangement for efficient cooling. Four cooling fans consume 360mW of power. The Arduino controls the cooling fan through the relay on/off operation. The solar panels had a 30W capacity for testing and were fixed at an optimal angle of 35 degrees, to maximize the solar energy. In addition, an infrared temperature sensor was attached to the back of the solar panel to monitor the temperature of the solar panel in real time. Figure 4 shows the four cooling fans attached to the solar panel.

Figure 4. The back of the solar panel with four cooling fans

2.3. Smart phone application part

This system does not allow people to go directly to the solar panel to turn the cooling fan on or off. The cooling fan should be controllable at anytime and anywhere. To implement this, control using smart phone application is essential. Open source software, called App Inventor 2, was used to develop apps. This is not only easy to make, but also available for free. An essential element to this application is the part that can monitor the sensor values received from the Arduino and the button part to control the cooling fan. Figure 5 presents the designed application and the source used in App Inventor 2.

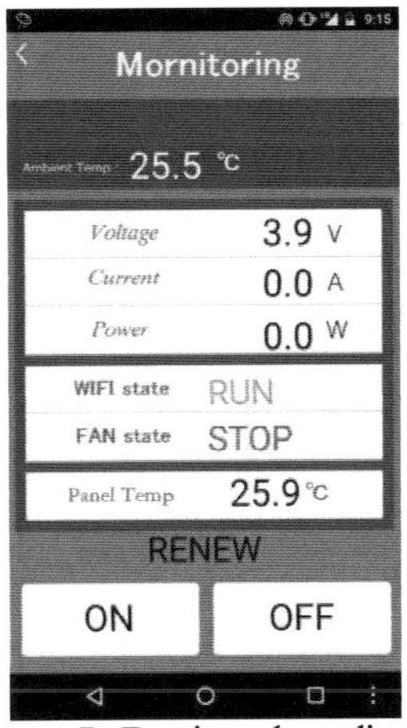

Figure 5. Designed application

3. Experimental Setup

Each component was designed and assembled to produce a model for the experiment. The experiment was carried out from 09:00 to 18:00 in clear, cloudless weather. The location was Pusan National University, Busan, Korea (31.040 N, 31.486 E), building number 207 on the roof, and the experiment date was around September 2017. The angle of the solar panel was maintained at 35 degrees, and the direction of gaze pointed to the south. An infrared thermometer was used to measure the temperature of the solar panel. A VT04A visual IR thermometer (FLUKE, Everett, WA, USA) was used. Model A without a cooling fan was compared with model B with a cooling fan. The system connected to the network for monitoring and remote control of the cooling fan. Figure 6 shows the layout.

Figure 6. Arranged solar panels (Model A on the left, Model B on the right)

4. Results and Discussion

The solar panel was fixed in a suitable wide area and the experiment was then conducted. At the beginning of the experiment (09:00 ~ 10:00) and at the end (16:00 ~ 17:00), the output power was similar because the temperature difference between the two solar panels was small. On the other hand, from 11:00 to 15:00, when the temperature difference of the solar panel is large, the solar panel with the cooling fan showed a higher output power than the solar panel without the cooling fan. The difference in the output power reached a maximum at 13:00 when the temperature difference in the solar panel reached its maximum. Figure 7 (a), (b), and (c) shows the solar panel temperature, ambient temperature, and output power over time.

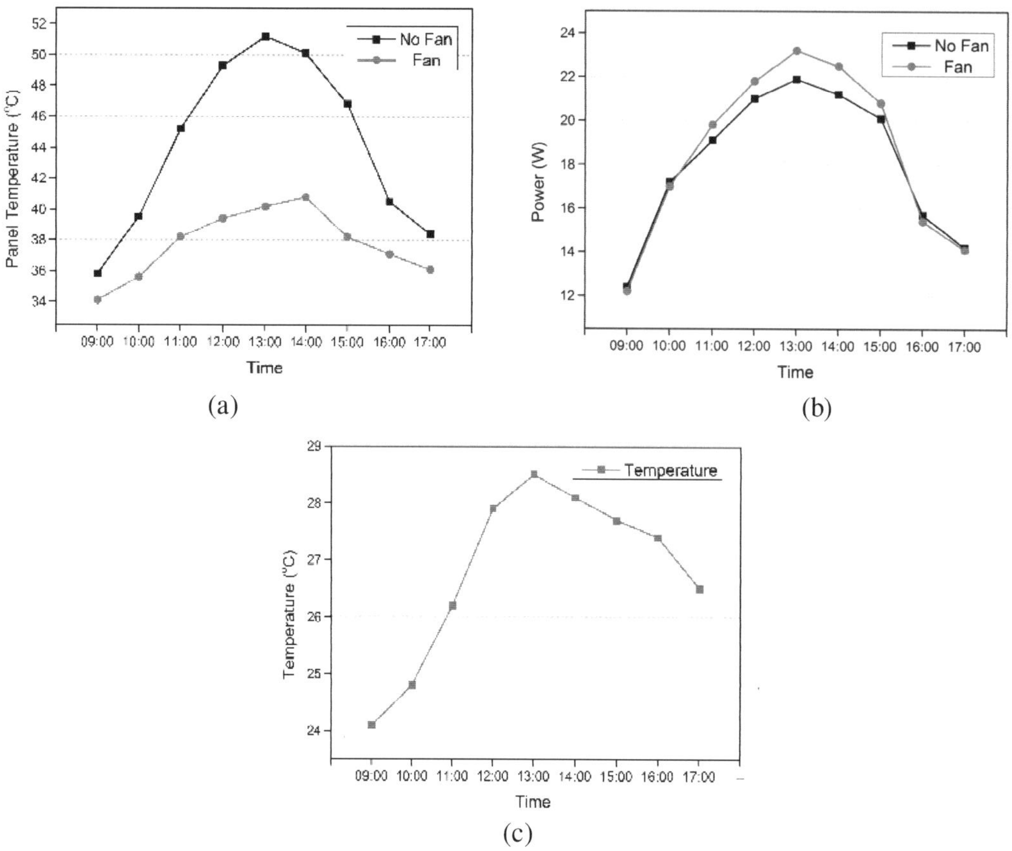

Figure 7. (a) Solar panel temperature (b) Output power of solar panel (c) Ambient temperature

The output power graph shows that the output power can be determined using this system from 11:00 to 15:00. At that time, the increase in efficiency of the output power was approximately 4.7%. This value can change depending on the season and weather. On the other hand, it is practical to use the IoT system to control at any time and place. Moreover, a higher power may be obtained by applying it to a large-capacity PV system rather than to such a small-capacity PV system.

5. Conclusion

The solar PV system becomes less efficient as the temperature of the solar panel increases. In particular, the temperature of the solar panel increases by more than 40 degrees during the daytime (11:00 ~ 15:00). Therefore, a cooling fan was installed on the back of the solar panel to increase the output power. On the other hand, when weather and other incidents occur, the solar system efficiency becomes low. If the temperature can be checked in real time and the cooling fan can be controlled automatically, the efficiency can be improved again. Therefore, real time remote control through an IoT system was implemented to increase the solar power efficiency. The IoT system device used is a microprocessor called Arduino, which makes it possible to construct an inexpensive and simple system. This system is compatible with various sensors and can implement various functions. Finally, after designing the IoT system and solar system, a successful experiment was conducted. Using a 30W solar panel, the maximum increase in efficiency was approximately 4.7% and the power increased by 1.4W. As a result, the performance improvement of the cooling fan using IoT was confirmed. If the system is also applied to a large-capacity PV system, higher power affected by the cooling fans can be expected. Future studies will extend the system and apply it to a 1kW PV system with ESS based on supercapacitor.

6. Reference

[1] Dubey, S.; Tiwari, G.N. Thermal modeling of a combined system of photovoltaic thermal (PV/T) solar water heater. Sol. Energy 2008, 82, 602–612.

[2] Cengel, Y.A.; Boles, M.A.Thermodynamics: An Engineering Approach, 5thed.; McGraw-Hill Science: New York, NY, USA, 2005.

[3] AMECIT. Informe de Smart Cities 2012. Available online: http://ametic.es/es/publicaciones/smart-tics (accessed on 5 July 2017).

[4] Doukas, C. Building Internet of Things with the Arduino, 1st ed.; CreateSpace: North Charleston, SC, USA, 2012; pp. 1–352.

[5] Kuniavsky, M. SmartThings: Ubiquitous Computing User Experience Design, 1st ed.; James, M., Ed.; Elsevier: San Francisco, CA, USA, 2010; pp. 1–287.

Application of the New Generation Large Capacity Synchronous Condenser in HVDC System

Zhilin Guo[1], Liangliang Hao[1] and Junyong Wu[1]

[1] School of Electrical Engineering, Beijing Jiaotong University, Beijing 100044 China

18117006@bjtu.edu.cn

Abstract. The AC dynamic reactive power compensation in high voltage direct current (HVDC) still remains an unsolved problem. Owing to its flexible and rapid reactive power output, synchronous condenser (SC) has a broad application prospect on sending and receiving ends of HVDC transmission system. Firstly, the system structure of a new large capacity SC is introduced. Secondly, reactive power dynamic characteristics of SC in sub-transient and transient are studied theoretically. Then its supporting effect on low bus voltage is analyzed and verified. On this basis, the reactive power application requirements of SC in DC system are studied. An actual transmission system model is established by PSCAD/EMTDC. The simulation verifies the suppression effect of SC on commutation failure and bus transient overvoltage, which provides a theoretical basis for the application of SC in HVDC system.

1. Introduction

During the 13th Five-Year Plan period, HVDC developed rapidly. Capacity of resource allocation in Chinese power system has been improved significantly. China's power grid tends to have a tendency of "strong DC system and weak AC system"[1]. With the rapid development of HVDC in interconnected power grids, lag in AC system has led to the issue that how to ensure the voltage stability through reactive power regulation in power grids. Five kinds of countermeasures have been put forward by power grid companies and academic institutions, such as strengthening dynamic reactive power construction, improving voltage control ability. etc. Compared with static var compensation(SVC), which has slow response speed and can't adjust the reactive power output , dynamic reactive power compensation can provide the reactive power needed by system under disturbances. It has a broad application prospect in HVDC transmission system[2,3].

Synchronous condenser (SC) is a common kind of dynamic reactive power compensation equipment. The new generation large capacity SC has fast and flexible regulation of reactive power output, which can fully enhance the dynamic reactive reserve of the system. The new generation large capacity SC can solve various voltage stability problems， such as dynamic reactive power shortage in receiving end, short circuit capacity insufficiency of weak sending end, etc [4-7]. It has a good response to new energy high voltage off-grid and commutation failure in HVDC transmission system [8-11].

Application analysis of SC in HVDC system has attracted attention. Reference [12-15] compare the reactive power output of SC with other dynamic reactive power devices in dynamic process and analyze the applicability of SC in HVDC; Reference [16-18] analyze the effect of parameter setting of SC on reactive power output.

Firstly, this paper introduces the structure of the new type large capacity SC system used in HVDC. Secondly, the sub-transient and transient effects of SC are analyzed. Then voltage regulation effect of SC during fault steady-state is analyzed and verified by analyzing the support effect on low voltage. On this basis, the application requirements for SC in HVDC are studied. Furthermore, the actual HVDC transmission line of Ximeng-Taizhou is built by PSCAD/EMTDC. The DC power and bus voltage under different working conditions are simulated. By comparing these waveforms with SC, its suppression effect on commutation failure and bus transient overvoltage is verified. This study provides a theoretical basis for the application of SC in HVDC transmission system.

2. Composition of the New Generation Large Capacity SC

SC is a common reactive power compensation device. It is an idle synchronous motor in essence. Since the last century, SC had been widely used in reactive power regulation of power grid. However, due to its limited regulating effect and high cost in operation and maintenance, SC had been gradually withdrew from operation. Unlike traditional salient pole structure, the new generation large capacity SC uses round rotor. In transient process, excitation voltage can reach 3.5 times, and excitation current can reach 2.5 times[16]. Typical parameters of the new generation SC are shown in Table 1. Compared with synchronous generators of the same capacity, the direct-axis super-transient reactance and the transient reactance of SC are significantly smaller, which makes it have a strong short-term excitation effect.

The composition of SC is shown in Figure 1. A self-excited structure is used by the new generation large capacity SC. The structure of AVR is shown in Figure 2. Its excitation control system is built according to the actual excitation system of the basic regulation and control function, including terminal voltage closed loop, system voltage loop, reactive power loop, constant rotor current closed loop which is used during field experiment and so on. It adopts PID and power system stabilizer (PSS) control mode with phase-in stator transient overcurrent limiting protection, phase-in reactive power lack-excitation limitation protection, load maximum instantaneous excitation current limiting protection, load maximum excitation current limiting protection and so on.

Table 1. Parameters of SC

Parameter	Unit	Value
rated capacity Q_1	Mvar	300
maximum allowable phase-in depth Q_2	Mvr	-150
nominal line voltage U	kV	20
rated current I	kA	8.66
unload excitation voltage U_{f0}	V	143
unload excitation current I_{f0}	A	735
d-axis synchronous reactance X_d	%	153
d-axis transient reactance X_d'	%	16.5
d-axis subtransient reactance X_d''	%	11.1
d-axis transient open-circuit time constant T_{d0}'	s	7.44
d-axis subtransient open-circuit time constant T_{d0}''	s	0.05
q-axis transient open-circuit time constant T_{q0}'	s	0.83
q-axis subtransient open-circuit time constant T_{q0}''	s	0.096

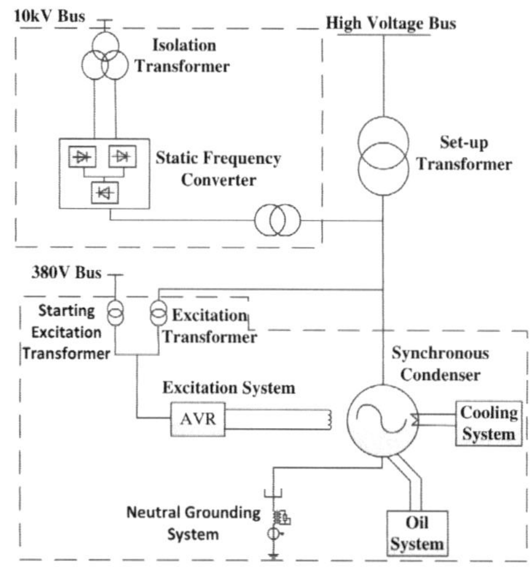

Figure 1. Composition of SC.

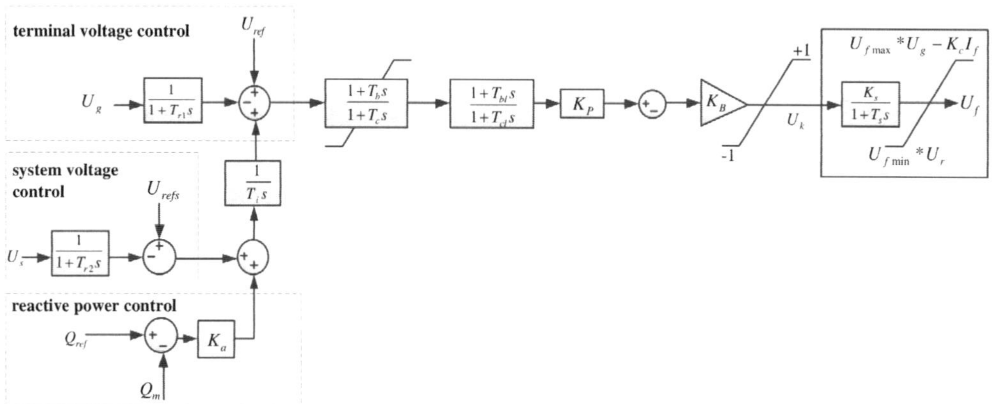

Figure 2. Structure of AVR

Where: U_{ref} is reference value of terminal Voltage. U_g is measurement value of terminal Voltage. U_{refs} is reference value of system Voltage. U_s is measurement value of system Voltage. Q_{ref} is reference value of reactive power. Q_m is measurement value of reactive power. U_f is field voltage. I_f is field current. K_a K_P K_B are gain coefficients. T_{r1} T_{r2} T_b T_c T_{bl} T_{cl} are time constants.

3. Dynamic Process and Demand Analysis of SC's Reactive Power

3.1. Theoretical Analysis of Reactive Power Dynamic Process of SC

Similar to generator, when system disturbances or faults occurs, SC also has subtransient and transient processes. Analyzing the dynamic process and the transient process of SC when the bus voltage of the system changes from the theoretical level is helpful to understand the effect of SC on reactive power regulation of the system.

Considering the difference between d axis and q axis, the phase diagram of SC is shown in Figure 3 [17].

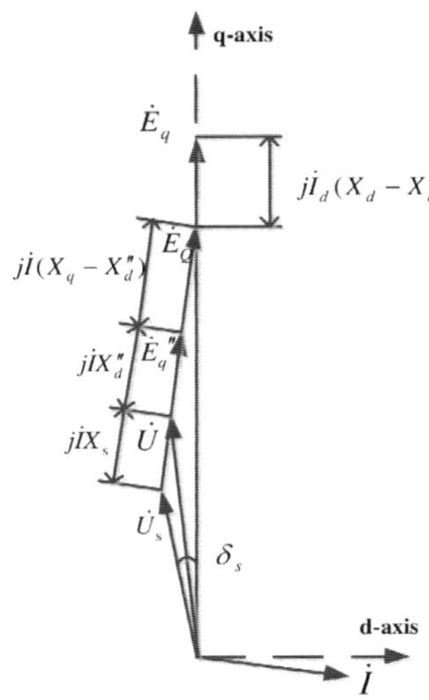

Figure 3. Phase diagram of SC.

In the figure, $\dot{U}_s$ is the high voltage of set-up transformer, $\dot{U}$ is the terminal voltage, $\dot{I}$ is the terminal current, $\dot{E}''_q$ is the q-axis sub-transient voltage, $\dot{E}_Q$ is the q-axis virtual electromotive force, $\dot{E}_q$ is the q-axis induction electromotive force, $\dot{I}_d$ is the d-axis component of the stator current. X_s is the contact reactance, X_d is the d-reactance, X''_d is the d-axis subtransient reactance, X_q is the q-axis reactance, $\delta_s \approx 0$ is the angle between $\dot{U}_s$ and $\dot{E}_q$.

The essence of SC is an idle motor, ignoring the stator resistance, there is $X_q \approx X_d$ for the round rotor, the reactive power of SC can be described as follow:

$$Q = u_q i_d - u_d i_q \approx u_q i_d \approx U_s i_d \tag{1}$$

Where: u_d, i_d are the d-axis voltage and current components, $u_d = U_s \sin\delta_s \approx 0$. u_q and i_q are the q-axis voltage and current components, $u_q = U_s \cos\delta_s \approx U_s$.

When bus voltage sags occur due to symmetrical fault, the variation of SC reactive power can be expressed as

$$\Delta Q = Q - Q_0 \approx U_s i_d - U_{s0} i_{d0} \approx (U_{s0} + \Delta U_s)(i_{d0} + \Delta i_d) - U_{s0} i_{d0} = U_s \Delta i_d + \Delta U_s i_{d0} \tag{2}$$

Where: Q is the postfault reactive power, Q_0 is the initial value of reactive power, i_{d0} is the initial value of d-axis current, Δi_d is the incremental of d-axis current, $\Delta U_s = U_s - U_{s0}$ is the variation bus voltage, U_s is the postfault bus voltage , U_{s0} is the initial value of the bus voltage. It can be seen from equation (2) that the variation of SC reactive power is related to the initial value and variation both bus voltage and stator current. In the same initial state, the greater the bus voltage drop, the greater the reactive power increment.

In the subtransient process, the excitation effect is neglected and only damper winding effect is considered. The d-axis stator current increment is showed as follows [17,18]:

$$\Delta i_d = \Delta U_s \left[-\left(\frac{1}{X''_d} - \frac{1}{X'_d} \right) e^{-\frac{t}{T''_d}} - \left(\frac{1}{X'_d} - \frac{1}{X_d} \right) e^{-\frac{t}{T'_d}} + \frac{\cos(\omega t + \delta)}{X''_d} e^{-\frac{t}{T_a}} - \frac{1}{X_d} \right] \tag{3}$$

Where: Δu_q is the variation of u_q, X'_d is the transient reactance of the d-axis, $T''_{ds} = T''_d \dfrac{X''_d + X_s}{X''_d}$ is the d-axis transient short-circuit time constant considered X_s. T''_d is the d-axis subtransient short-circuit time constant of SC, $T'_{ds} = T'_d \dfrac{X'_d + X_s}{X'_d}$ is the d-axis transient short-circuit time constant considered X_s, T'_d is the d-axis transient short-circuit time constant.

It can be known from equation (3) that the instantaneous increment of the reactive current is $\Delta i_d = -\dfrac{\Delta U_s}{X''_d + X_s}$, and the inductive reactive power provided by SC to the system in case of power grid failure is obtained by equation (2):

$$\Delta Q = -\frac{U_s \Delta U_s}{X''_d + X_s} + \Delta U_s i_{d0} = -\Delta U_s \left(\frac{U_s}{X''_d + X_s} - i_{d0}\right) \tag{4}$$

According to equation (4), the reactive power support of SC is positively correlated with $-\Delta U_s$ during the subtransient process, that is, when the bus voltage rises($\Delta U_s > 0$), the reactive power output decreases($\Delta Q < 0$), and the reactive power output increases ($\Delta Q > 0$)when the bus voltage decreases ($\Delta U_s < 0$). The reactive power of SC $Q_0 \approx 0$ during normal operation, and from equation (1), it is known that $i_{d0} \approx 0$. From equation (4) , $\Delta Q > 0$ when bus voltage drops and $\Delta Q < 0$ when bus voltage rises, that is, SC absorbs reactive power when the bus voltage rises, and emits reactive power when the bus voltage decreases.

The influence of damper winding and stator transient process are not considered in transient process. The d-axis stator current increment in transient process can be expressed as:

$$\Delta i_d = -\frac{\left[K_{3E}(K_A + 1) - 1\right]\Delta U_s}{(K_{3E} T'_{d0} s + 1)(X'_d + X_s)} - \frac{\Delta U_s}{(X'_d + X_s)} \tag{5}$$

Where: T'_{d0} is the d-axis open-circuit transient time constant, X'_d is the d-axis transient reactance, $K_{3E} = \dfrac{(X'_d + X_s)}{X_d + X_s + K_A X_s}$, K_A is the excitation amplification factor.

From equation (2) and equation (5), we can know the reactive power increment generated by SC during transient process:

$$\Delta Q = -\frac{\left[K_{3E}(K_A + 1) - 1\right]U_s}{(K_{3E} T'_{d0} s + 1)(X'_d + X_s)}\Delta U_s - \frac{U_s}{(X'_d + X_s)}\Delta U_s + \Delta U_s i_{d0} = -\Delta U_s \left(\frac{\left[K_{3E}(K_A + 1) - 1\right]U_s}{(K_{3E} T'_{d0} s + 1)(X'_d + X_s)} + \frac{U_s}{(X'_d + X_s)} - i_{d0}\right) \tag{6}$$

It can be seen from equation (6) that, similar to subtransient effect, the variation of the reactive power and the bus voltage show a reverse change trend in transient process.

In summary, when the terminal voltage decreases or rises, SC can adjust the reactive power immediately under the subtransient process, and then continuously emit or absorb reactive power during transient process under the excitation, which is conducive to the restoration of bus voltage and the reduction of fault impact. The sub-transient and transient effects of SC are also the fundamental reason for its flexible and fast reactive power output.

3.2. Analysis of the Supporting Effect of SC on Low Voltage

Taking the supporting effect of SC on transient low voltage as an example, the regulating effect of SC on the bus voltage during fault steady state is analyzed. Regardless of the influence of DC system and filter, the system model is shown in Figure 4. Ignoring the restriction of self-shunt excitation, when SC reaches the maximum excitation effect (the terminal voltage reaches the pre-fault level with excitation controller during the fault), the effect of bus voltage lifting is the greatest. The effect of SC on low bus voltage during the fault is analyzed.

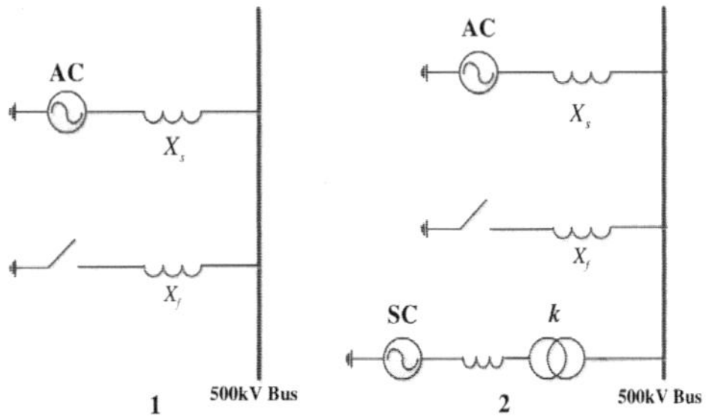

Figure 4. System model with and without SC.

The model of system 1,2 is shown in Figure 4. In the model, AC is the equivalent alternating current power supply connected by bus, V_s is the internal potential of AC power supply, X_s is the connection reactance between AC system and the bus, and a symmetrical faults occur at the bus (transition reactance is X_f). Model 1 does not have SC. One SC is connected to the bus by the main transformer (leakage reactance is X_T, ratio is k) in model 2. The model parameters are as follows: $V_s = 560\text{kV}$, $V_{sc} = 20\text{kV}$, $X_s = 10.1662\Omega$, $Z_f = 19.792\Omega$, $X_T = 0.1333\Omega, k = 20/530$

Bus voltage per unit value: $V_B = 530\text{kV}$

The steady-state postfault bus voltage:

$$\text{Model 1:} V_1 = \frac{X_s}{X_f + X_s} V_s \tag{7}$$

$$\text{Model 2:} V_2 = X_f \left(\frac{V_s - V_2}{X_s} + k \frac{V_{sc} - kV_2}{X_T} \right) = \frac{V_s + kV_{sc} X_s (X_T)^{-1}}{X_s (X_f)^{-1} + k^2 X_s (X_T)^{-1}} \tag{8}$$

The increase of bus voltage caused by SC:

$$\Delta V = V_2 - V_1 = \frac{V_s + kV_{sc} X_s (X_T)^{-1}}{X_s (X_f)^{-1} + k^2 X_s (X_T)^{-1}} - \frac{X_s}{X_f + X_s} V_s = \frac{V_{sc} k^{-1} - V_s \left(X_s B_f + 1 \right)^{-1}}{\left(B_f + X_s^{-1} \right) X_T k^{-2} + 1} = 10.53655\text{kV} \ (0.01988.\text{p.u}) \tag{9}$$

Where: $B_f = X_f^{-1}$

The model shown in Figure 4 is built by PSCAD. Under the same AC level and fault conditions, the influence of SC is simulated. The results are shown in Figure 5.

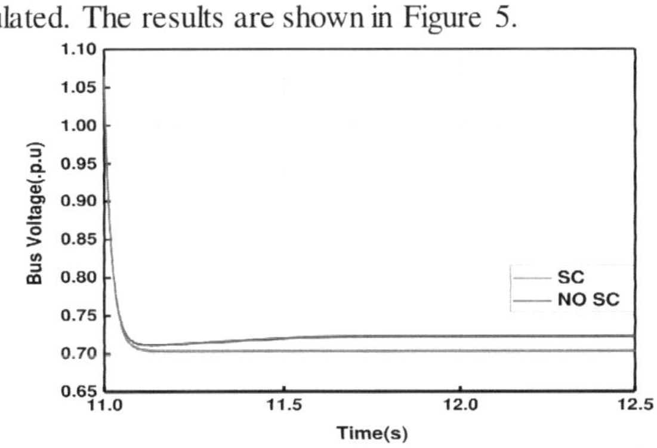

Figure 5. The bus voltage with and without SC

From Figure 5, $\Delta V = 0.0197.\text{p.u}$ at the steady-state during the fault. Considering the simulation measurement error and other factors, the simulation results are basically consistent with the theoretical calculation results. The low bus voltage can be improved by SC.

3.3. Application of SC in HVDC

The reactive power of HVDC system is mainly supplied by SVC devices such as filters and series capacitors during normal operation. The capacity of SVC is sufficient. When the system is in normal operation or has small disturbances, the basic reactive power requirement can be met by switching on and off filter banks. SC serves as reactive backup for the system and it's reactive power output is basically zero. Equation (10) shows that the reactive power output of SVC in the system has positive fluctuation with bus voltage fluctuation, which goes against the recovery bus transient voltage.

$$Q_c = \frac{U_c^2}{Z_c} \tag{10}$$

Where: Q_c is the reactive power output of filter, U_c is the terminal voltage of filter and Z_c is the impedance value of filter.

The main cause of commutation failure is the low voltage caused by the fault of AC bus on receiving end of HVDC transmission system. The continuous commutation failure of HVDC system may even lead to DC blocking and the collapse of system. When bus voltage decreases, synchronous condenser on the receiving end can give full play to the short-term strong excitation and provide certain reactive power support, improve the transient low voltage, effectively avoid commutation failure.

Since sending end of HVDC contains a large amount of new energy, the over-voltage of sending end easily causes new energy high voltage trip-off, which seriously threatens the safe and stable operation of the system. DC restart and blocking can cause the overvoltage. Taking DC blocking as an example, the delayed removal of SVC causes a large number of reactive surplus and bus overvoltage in the near region when bipolar blocking fault occurs in the system. At this time, synchronous condenser enters the phase-leading operation spontaneously to absorb excess reactive power and restrain bus overvoltage.

4. Simulation analysis

4.1. Introduction of simulation system

In order to verify the role of synchronous condenser in HVDC system, the traditional HVDC transmission model is built by PSCAD/EMTDC. Based on the actual HVDC project in Ximeng-Taizhou, the model adopts a bipolar 12-pulse valve series structure with a rated power of 10,000 MW and a rated voltage of 800 kV [19,20]. Ximeng is the sending end and AC voltage is 500 kV. One synchronous condenser is settled on sending end. Taizhou is the receiving end of the DC system, which is connected to 500 kV and 1000 kV AC network in layers. The system consists of converter, converter transformer, AC filter (ACF), DC filter (DCF), smoothing reactor, grounding electrode and so on. The topology structure is shown in Figure 6 [19, 20].

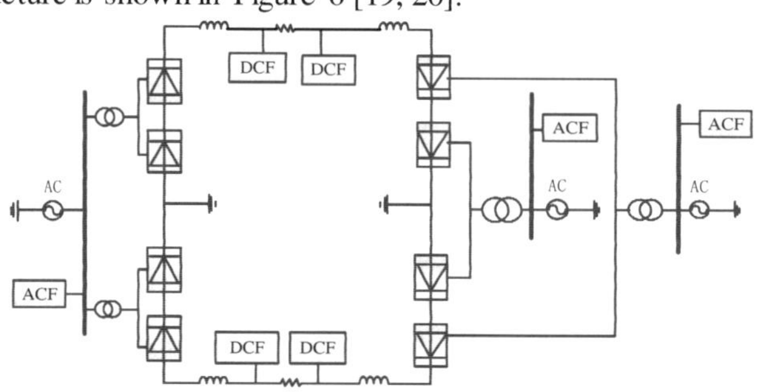

Figure 6. Structure of Ximeng-Taizhou HVDC system

The sending end control system includes constant current control which can maintain the value of DC current equivalent to its reference, minimum α control which keeps $\alpha \geq 5^{\circ}$ in order to prevent the shutting of converter. The inverter side includes constant γ control which can maintain the value of γ equivalent to its reference and constant current control which is used during fault time. In addition, there are low-voltage current limiting(When voltage decreases, current reference value is reduced to reduce the system's demand for reactive power) and current deviation control which can realize the steady transition from constant γ control to constant current control in the transient process. In normal operation, constant current control of sending end and constant γ control of receiving end work together to make the system run stably under rated conditions. Sending end is switched to minimum α angle control and receiving end is switched to constant current control during sending end fault; when the fault occurs on receiving end, sending end is the constant current control with the action of the low-voltage current limiting module. The inverter side is constant γ control.

In order to verify the restraint effect of synchronous condenser on commutation failure and bus overvoltage in HVDC transmission system, AC fault at DC receiving bus and DC blocking fault of HVDC transmission system are respectively set up by PSCAD. Taking three-phase grounding fault as an example, fault occurs at the receiving bus, which leads to commutation failure at the receiving end of HVDC transmission system. By measuring the electrical quantities of the system during fault period, the influence of synchronous condenser on DC system is proved.

4.2. The Effect of SC on Commutation Failure

A symmetrical faults (fault time t=11s) occurs at 75.6km away from the 500 kV bus on receiving end. Considering of the main protection action time of 500kV AC line at the far end, the fault lasts for 100ms. In order to verify the effect of SC on commutation failure, DC power, receiving end bus voltage and turn-off angle are measured. The most essential reason for the commutation failure of traditional HVDC transmission is that its converter valve is SCR. When the time between its turn-off to positive zero-crossing point of the voltage is less than the limit turn-off angle γ_{min} ($\gamma_{min} \approx 8°$), the carrier recombination of converter valve will lead to the conduction of converter valve without trigger pulse. Turn-off angle less than γ_{min} is the standard for commutation failure. The simulation results are shown in Figure 7.

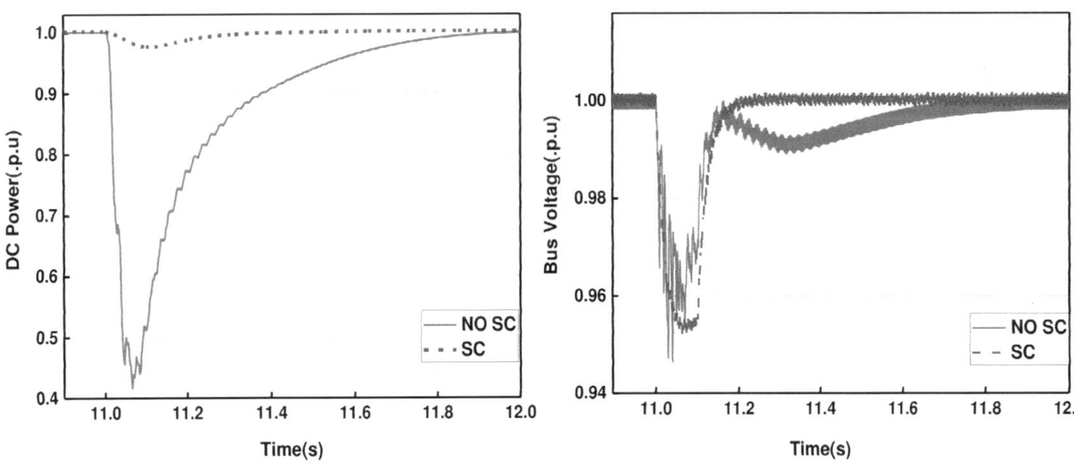

(a) (b)

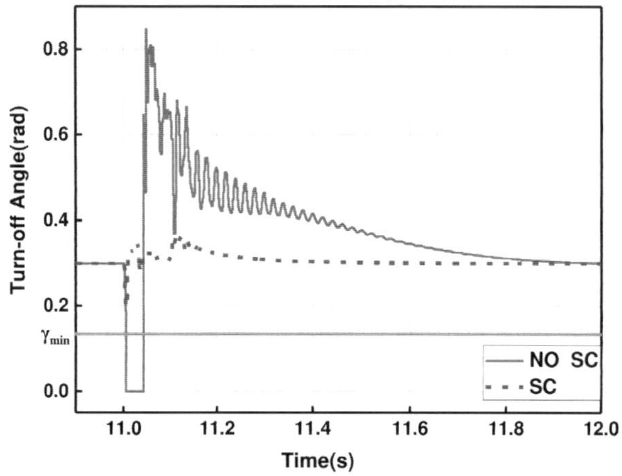

(c)

Figure 7. Simulation results of AC fault on receiving end.

It can be seen from Figure 7 that during AC failure on receiving end, the bus voltage drops instantaneously when SC is not connected, at this time receiving end turn-off angle is less than γ_{min} and DC power is greatly reduced, commutation failure occurs on receiving end. When receiving end is connected to a SC, the transient voltage decrement is reduced after the fault, the turn-off angle is always greater than γ_{min} during the fault and the power reduction is also significantly smaller than that without SC. There is no commutation failure in the system with a SC connected to receiving end. In the simulation, for same fault, DC system with SC installed on the receiving end can resist the influence of the AC fault and avoid the occurrence of the commutation failure.

4.3. The suppressive effect of SC on overvoltage

DC blocking occurs in DC system (fault time t=11s). DC power, bus voltage of sending end and reactive power output of SC are used to give proof of the restraint effect of SC on bus overvoltage. The simulation results are shown in Figure 8.

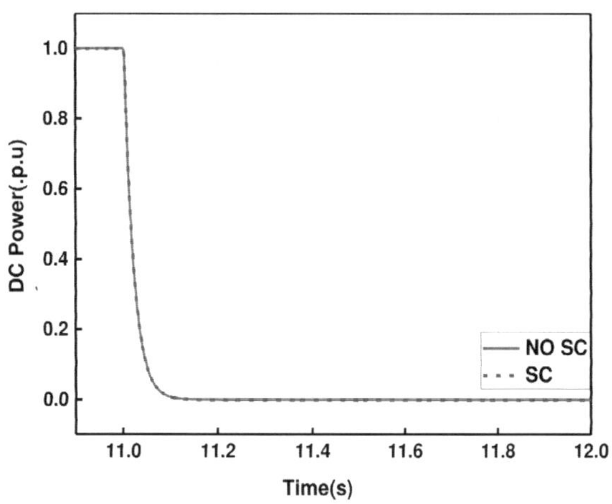

(a)

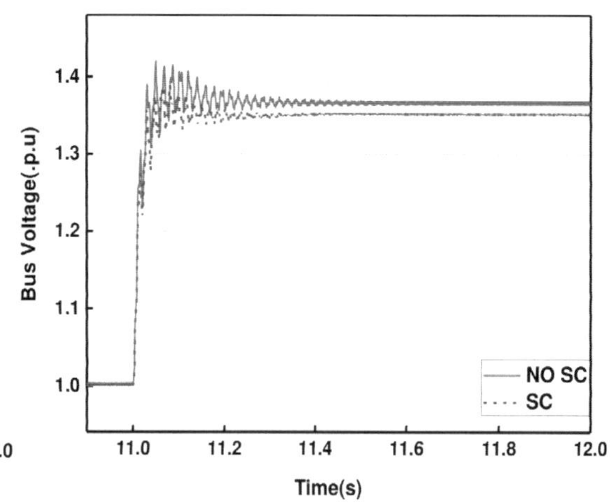

(b)

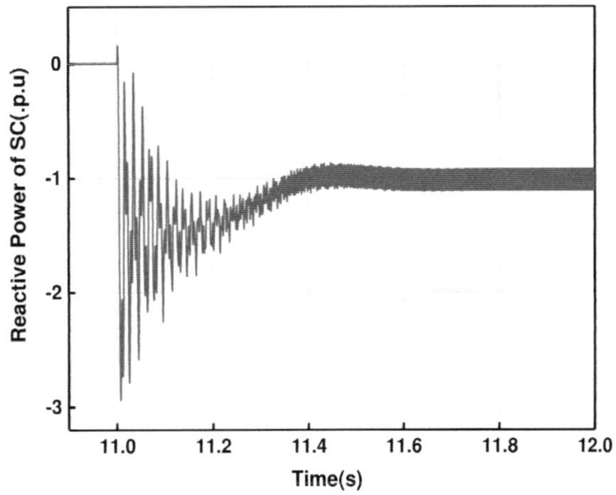

(c)

Figure 8. Simulation results of DC blocking.

As can be seen from Figure 8, the DC power rapidly drops to zero after DC blocking. As a result, whether SC is connected or not has no influence on DC power. Sending end bus voltage rises rapidly, transient over-voltage can raise to 1.42.pu without SC. The maximum transient over-voltage drops by 0.04.pu during the fault with SC. It quickly absorbs reactive power and the maximum up to 2.97.pu under the subtransient process, which can also be seen from Fig7(c). The fault transient overvoltage level is effectively suppressed by the access of SC on sending end. Since sending end contains a large amount of new energy, the overvoltage of the bus voltage may cause a series of chain problems such as high voltage off-grid of new energy source, and the connection of sending end to SC can avoid the above problems to a certain extent.

In summary, it can be seen that the instantaneously flexible reactive power of SC has a certain suppression effect on the low voltage and overvoltage of the bus. The SC applied to DC system sending and receiving end have good countermeasures to solve commutation failure and new energy high voltage off-grid.

5. Conclusion

This paper analyses the application of the new generation large capacity synchronous condenser in HVDC system. Firstly, the synchronous condenser's system structure is introduced. Then the dynamic varying process of reactive power along with terminal voltage is analyzed theoretically. Taking the supporting effect on the low voltage as an example, the regulating effect of synchronous condenser on the bus voltage during the steady state of the fault is calculated and verified. Moreover, an actual HVDC transmission is used as the model to simulate the waveform changes of each electrical quantities before and after the synchronous condenser is connected. Conclusions are as follows:

(1) When the terminal voltage changes, synchronous condenser can adjust the reactive power output flexibly immediately during the sub-transient and transient process, and the reactive power output changes with the terminal voltage inversely, which is conducive to the recovery of transient voltage.

(2) The new generation large capacity synchronous condenser has an effect of supporting low bus voltage. Occurrence of commutation failure on receiving end of DC system can be reduced with the application of synchronous condenser.

(3) Large capacity synchronous condenser has an effect of restraining bus overvoltage. The risk of high voltage off-grid of new energy can be reduced to a certain extent after the application of synchronous condenser, thus providing a guarantee for the safe and stable operation of DC system.

6. References

[1] WEN Jialiang, WU Rui, PENG Chang, et al. Analysis of DC Grid Prospects in China[J]. *Proceedings of the CSEE*,2012,**32**(13): 7-12+185.

[2] LIU Zhenya, ZHANG Qiping, WANG Yating. Research on Reactive Compensation Strategies

for Improving Stability Level of Sending-end of 750kV Grid in Northwest China [J]. *Proceedings of the CSEE*, 2015(**05**):1015-1022.

[3] WANG Yating, ZHANG Yichi, ZHOU Qinyong, et al. Study on application of new generation large capacity synchronous condenser in power grid [J]. *Power System Technology*, 2017, **41**(01):22-28.

[4] TU Jingzhe, ZHANG Jian, LIU Mingsong, et al. Study on Wind Turbine Generators Tripping Caused by HVDC Contingencies of Wind-Thermal-Bundled HVDC Transmission [J]. *Power System Technology*, 2015, **39**(12):3333-3338.

[5] HAN Pingping, ZHANG Haitian, DING Ming, et al.A Coordinated HVRT Strategy of Large-Scale Wind Power Transmitted With HVDC System[J].*Power System Technology*, 2018(**04**):1086-1095.

[6] ZHU Pengfei, YAN Naichen, ZHANG Chong, et al. Process Analysis of Reactive Power Output of 300MVar New Air Cooled Condenser Phase Modifier in Bipolar Blocking Process[J]. *Electronic Components and Information Technology*, 2018, **2**(06):111-115.

[7] FU Min, HE Yingming, YIN Zhaoyang. Analysis of Performance of Reactive Power Compensators in Weak Sending System in HVDC[J/OL]. *Journal of North China Electric Power University (Natural Science Edition)*:1-8[2018-12-19].

[8] HAO Xiaohong, GUO Tie e, WANG Ningbo, et al. Research on Application of Large-capacity Synchronous Condenser in Hexi Sending Terminal Hybrid Power Grid[J]. *Journal of Gansu Sciences*, 2018, **30**(5):113-118.

[9] XU Pengjian, HUANG Jinhai, XU Jingjing, et al. Study of Bin-Jin UHVDC project commutation failures protection[J]. *Power System Protection and Control*, 2017, **45**(02):140-146.

[10] He Chaorong, Li xingyuan, Jin xiaoming, et al. Simulation analysis on commutation failure criteria for HVDC transmission systems. *Power System Technology*, 2007(**01**):20-24.

[11] WANG Qing, SHA Jiangbo, YANG Pengpeng, et al. Study of impact of SC on the ability to defend commutation failure of LCC-HVDC[J].*Advanced Technology of Electrical Engineering and Energy*,2018,**37**(05):29-36.

[12] TELEKE S, ABDULAHOVIC T, THIRINGER T, et al. Dynamic Performance Comparison of Synchronous Condenser and SVC[J]. *IEEE Transactions on Power Delivery*, 2008,**23**(3):1606-1612.

[13] JIN Yiding,YU Zhao, LI Mingjie, et al. Comparison of New Generation Synchronous Condenser and Power Electronic Reactive-Power Compensation Devices in Application in UHV DC/AC Grid[J]. *Power System Technology*, 2018, **42**(07):2095-2102.

[14] KYNEV S, PILZ G, SCHMITT H. Comparison of modern STATCOM and synchronous condenser for power transmission systems, 2016[C]. *IEEE*, 2016.

[15] XU Guorui, FAN Shixiong, LUO Chaolong, et al. Operation Characteristic of Synchronous Condenser and SVC for Dynamic Reactive Power Compensation of DC Converter Station[J/OL].*Journal of North China Electric Power University (Natural Science Edition)* :1-7[2018-12-19]

[16] LI Zhiqiang, JIANG Weiyong, WANG Yanbin, et al. Key Technical Parameters and Optimal Design of New Types of Large Capacity Synchronous Condenser[J]. *Large Electric Machine and Hydraulic Turbine*, 2017(**04**):15-22.

[17] GUO Yibing, LING Zaixun, CUI Yibo,et al. Characteristics analysis of large synchronous condenser for UHV DC project[J]. *HUBEI ELECTRIC POWER*, 2016,**40**(05):1- 4+34.

[18] GAO Jingde, WANG Xianghang, LI Fahai. *Analysis of AC electrical machine systems*[M]. Beijing: Tsinghua University Press, 1994:166-193.

[19] LI Gaowang, LI Yanan, ZOU Xin.AC-Side transient overvoltage in ± 800 kV Ximeng-Taizhou UHVDC transmission project[J].*Electric Power Construction*, 2015,**36**(09):57-61.

[20] LI Bo, SU Yi-jing, ZHAO Bin, et al. Study and design of main circuit parameters for ± 800 kV /10000 MW DC power transmission project from Ximeng to Taizhou[J]. *Energy Engineering*,2017(**03**):27-33.

IOP Conf. Series: Earth and Environmental Science **342** (2019) 012008 doi:10.1088/1755-1315/342/1/012008

An Improved Droop Control Method for Energy Storage Module of DC Microgrid

J W Lin[1], Z X Wang[1*], L X Bao[2] and H Qin[3]

[1] School of Electronics, Information and Electrical Engineering, Shanghai Jiao Tong University, 800 Dongchuan Road, Shanghai 200240, China
[2] Shanghai Xilong Technology Co., Ltd, Shanghai 201517, China
[3] Shanghai Huacheng Elevator Technology Co., Ltd, Shanghai 201111, China

*Corresponding author: wangzxin@sjtu.edu.cn

Abstract. This paper proposes an improved droop control method with state of charge (SOC) balance function for energy storage module of DC microgrid. The average SOC of all energy storage modules is obtained by low bandwidth network. The SOC balance controller calculates the difference between each module's SOC and average SOC and then adjust the droop coefficients according to the difference. As the coefficient changes, the output power of module is allocated according to its SOC and eventually achieve balance. Besides, secondary voltage controller is employed to compensate DC bus voltage drop. Also, the stability of the proposed method is studied by small-signal analysis. The effectiveness of the proposed method is finally verified by MATLAB/Simulink simulation.

1. Introduction

With the continuous development of renewable energy, the concept of microgrid came into being[1]. Since DC microgrid does not have the problems of phase synchronization, reactive power loss, harmonics, etc. that often occur in ac microgrid, the research and application of DC microgrid has received more and more attention worldwide [2-4]. A DC microgrid system is shown in Figure 1. In order to stabilize the energy fluctuation of renewable energy, energy storage devices are usually added to the DC microgrid[5]. As implication, multiple energy storage modules are typically distributed in the microgrid to meet the needs of large-capacity systems [6, 7]. It is important for energy storage module to allocate the energy properly, which helps to prevent overdischarge and extend its life.

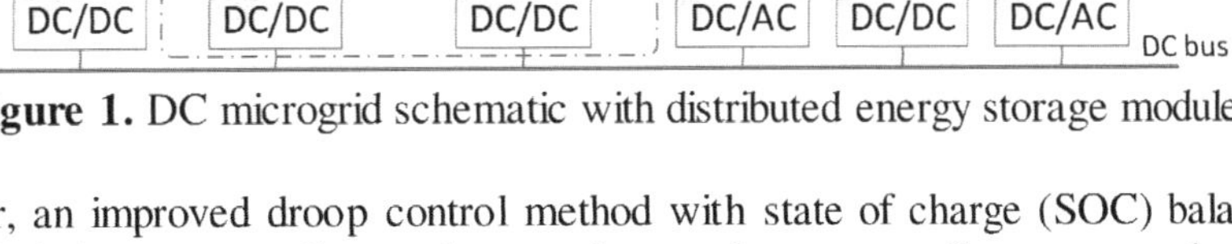

Figure 1. DC microgrid schematic with distributed energy storage modules

In this paper, an improved droop control method with state of charge (SOC) balance function is proposed. SOC balance controller and secondary voltage controller are employed, and SOC information is exchanged via low bandwidth network. The SOC balance controller is used to regulate the droop coefficient based on the difference between each parallel energy storage module's SOC and the average SOC, which adjusts the output power module so that the SOC tends to be equal. The secondary voltage controller compensates for bus voltage drop.

The rest of the paper is organized as follows. In section 2, the improved droop control is introduced and the stability of the method is studied with small-signal analysis. Section 3 shows the results and discussions of MATLAB/Simulink simulation. Section 4 finally summarizes the paper and draws the conclusions.

2. Principle of improved droop control method

Droop control is a common method for energy management in DC microgrid [8-10]. In this paper, we improve the traditional droop control by changing the droop coefficient to achieve SOC balance among different energy storage modules. The details of the method are discussed as follows.

2.1. SOC balance controller

The decrease rate of SOC is positively correlated with the output power. Consider two parallel energy storage modules with the same specification but different initial SOC. To equalize the SOC of two modules, the droop coefficient of the module with higher SOC should be increased, thus output power increases and SOC decreases faster. Whereas, the droop coefficient of the module with lower SOC should be decreased. When SOC of two modules are the same, the droop coefficients will remain unchanged so that the parallel modules maintain both SOC and output power balance. The modules' SOC are exchanged via low bandwidth network and the average SOC is calculated.

The adjustment of droop coefficient is expressed as follows:

$$m_i = m_0 - m_{i0}(SOC_i - SOC_{avg}) = m_0 - m_{i0}(SOC_i - \sum_{i=1}^{N}\frac{SOC_i}{N}) = m_0 - m_{i0}\Delta SOC_i \quad (1)$$

where m_i is the calculated droop coefficient of module i, m_0 is the given initial value of droop coefficient, m_{i0} is the scale factor, SOC_i is the SOC of module i, N is the number of all modules put into operation, SOC_{avg} is the average SOC of all N parallel modules.

It can be seen from (1) that, m_i will be smaller than m_0 if SOC_i is bigger than SOC_{avg} and conversely m_i will be bigger than m_0 if SOC_i is smaller than SOC_{avg}, which leads to increase or decrease in output power of each module. As ΔSOC_i gradually drops to zero, m_i tends to m_0 so that the output power of each module will not change anymore and the SOC is balanced.

m_{i0} can not be too large in case m_i is too small or even below zero, which will cause instability of the system. Normally, the values of m_{i0} are the same in all modules. It should also be noticed that both the parameters m_0 and m_i should be given approximately to ensure that the output power of each module is within limit.

2.2. Stability analysis of SOC balance controller

To analyse the stability of SOC balance controller, we take a DC microgrid with two parallel energy storage modules as an example without loss of generality. Assume that two modules have the same capacity and rated voltage but different SOC. The reference voltage of two modules can be express as

$$u_{dc1} = u_{dc}^* - [m_0 - m_{10}(soc_1 - soc_{avg})]p_1 \quad (2)$$

$$u_{dc2} = u_{dc}^* - [m_0 - m_{20}(soc_2 - soc_{avg})]p_2 \quad (3)$$

where u_{dc1} and u_{dc2} are the reference voltage, u_{dc}^* is the no-load voltage, soc_1 and soc_2 are the SOC from low bandwidth network, soc_{avg} is the average SOC of two modules, p_1 and p_2 are the measured output power.

Since soc_{avg} can be derived as

$$soc_{avg} = (soc_1 + soc_2)/2 = soc_1 - \Delta soc = soc_2 + \Delta soc \quad (4)$$

The Δsoc, p_1 and p_2 can be reached as

$$\Delta soc = (soc_1 - soc_2)/2 = G_d \Delta SOC = [1/(\tau s + 1)]\Delta SOC \quad (5)$$

$$p_1 = G_{lpf} P_1 = [\omega_c / (\omega_c + s)]P_1 \quad (6)$$

$$p_2 = G_{lpf} P_2 = [\omega_c / (\omega_c + s)] P_2 \tag{7}$$

where SOC_{avg} is the actual average SOC, P_1 and P_2 are the actual output power, G_d indicates the delay of low bandwidth network, τ is the time constant of delay link, G_{lpf} is the transfer function of lowpass filter, ω_c is the cut-off frequency of lowpass filter.

Substituting (4)~(7) into (2) and (3), it yields

$$u_{dc1} = u_{dc}^* - [m_0 - m_1 G_d \Delta SOC] G_{lpf} P_1 \tag{8}$$

$$u_{dc2} = u_{dc}^* - [m_0 + m_2 G_d \Delta SOC] G_{lpf} P_2 \tag{9}$$

Performing the small-signal analysis for (8) and (9), it yields

$$\hat{u}_{dc1} = (m_1 G_d \Delta SOC - m_0) G_{lpf} \hat{P}_1 + m_1 G_d G_{lpf} P_1 \Delta \hat{SOC} \tag{10}$$

$$\hat{u}_{dc2} = (-m_2 G_d \Delta SOC - m_0) G_{lpf} \hat{P}_2 - m_2 G_d G_{lpf} P_2 \Delta \hat{SOC} \tag{11}$$

At point of common coupling, the following equation is satisfied.

$$P_1 + P_2 = U_{dc}^2 / R_{load} \tag{12}$$

where U_{dc} is the DC bus voltage and R_{load} represents the load resistance.

Performing the small-signal analysis for (12), it yields

$$\hat{P}_1 + \hat{P}_2 = 2U_{dc} \hat{U}_{dc} / R_{load} \tag{13}$$

Because the response speed of the inner current loop is much faster than droop control, the closed loop transfer function of voltage can be considered as 1[11]. The approximate relation is shown as

$$U_{dc} \approx u_{dc1} \approx u_{dc2} \tag{14}$$

Combining (10), (11), (13) and (14), the following closed loop transfer functions are obtained

$$G_{\hat{U}_{dc}/\Delta\hat{SOC}}(s) = \hat{U}_{dc} / \Delta\hat{SOC} = a[\tau s^2 + (\omega_c \tau + 1)s + \omega_c] / (bs^3 + cs^2 + ds + e) \tag{15}$$

where $a = -m_1^2 \Delta SOC (P_1 + P_2) R$, $b = -2m_0 R\tau$, $c = -2m_0 \tau (2R + R\omega_c \tau - U_{dc} m_0 \omega_c \tau)$, $d = -2m_0 (R + 2R\omega_c \tau + 2U_{dc} m_0 \omega_c \tau)$, $e = -2\omega_c (m_0 R + U_{dc} m_0^2 - m_1^2 U_{dc} \Delta SOC^2)$.

$$G_{\hat{U}_{dc}/\hat{P}_1}(s) = \hat{U}_{dc} / \hat{P}_1 = (As + B) / (Cs^2 + Ds + E) \tag{16}$$

where $A = (P_1 - P_2) R m_0 \omega_c \tau$, $B = R\omega_c [m_0 (P_1 - P_2) + m_1 (P_1 + P_2) \Delta SOC]$, $C = (P_1 + P_2) R\tau$, $D = (P_1 + P_2)(R\omega_c \tau + 1) + 2P_1 U_{dc} m_0 \omega_c \tau$, $E = (P_1 + P_2) R\omega_c + 2P_1 U_{dc} \omega_c (m_0 + m_1 \Delta SOC)$.

Solve the dominant poles of (15) and (16) by using MATLAB and the results are shown in figure 2 and figure 3. When ΔSOC varies over a wide range, the dominant poles locate in the left half of s domain, which guarantees the stability of SOC balance controller.

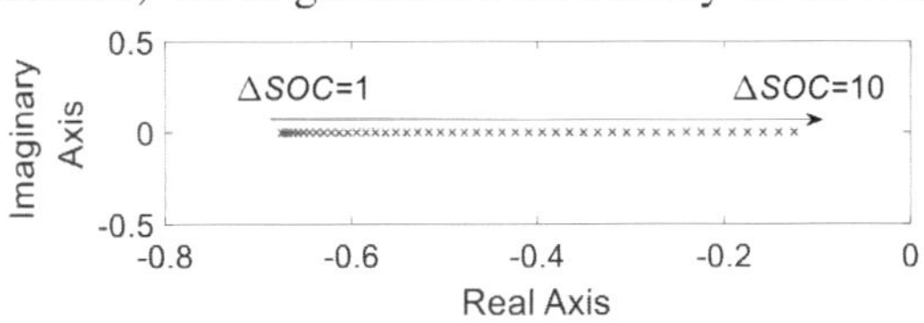

Figure 2. Dominant poles map of $G_{\hat{U}_{dc}/\Delta\hat{SOC}}(s)$

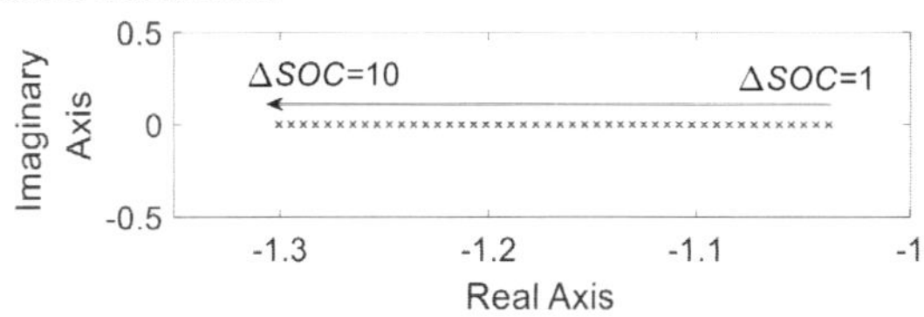

Figure 3. Dominant poles map of $G_{\hat{U}_{dc}/\hat{P}_1}(s)$

2.3. Secondary voltage controller

In this paper, secondary control is introduced to prevent voltage drop caused by droop control. The stability of the system needs to be verified. The control scheme is shown in Figure 4, the dotted line part indicates the droop control which can be regarded as a disturbance to the secondary control[12]. The secondary control diagram can be expressed as

$$u_{dci} = u_{dc}^* - [m_0 - m_1 G_d \Delta SOC_1] G_{lpf} P_1 + G_{pi} (u_{dc}^* - u_{dci}) \tag{17}$$

Rewrite (17) as

$$u_{dci} = u_{dc}^{*} - \frac{[m_0 - m_1 G_d \Delta SOC_1] G_{lpf}}{1 + G_{pi}} P_1 = u_{dc}^{*} - \frac{[m_0 - m_1 G_d \Delta SOC_1] G_{lpf}}{1 + (K_p + K_i / s)} \times P_1 \tag{18}$$

where G_{pi} is the PI control transfer function, K_p and K_i are proportion and integral coefficient.

Figure 5 shows the frequency response of secondary controller with the change of ΔSOC_1. As ΔSOC_1 changes over a wide range, the amplitude is far below zero which means that P_1 has little effect on the output of secondary controller. Hence, the stability of secondary voltage controller is guaranteed.

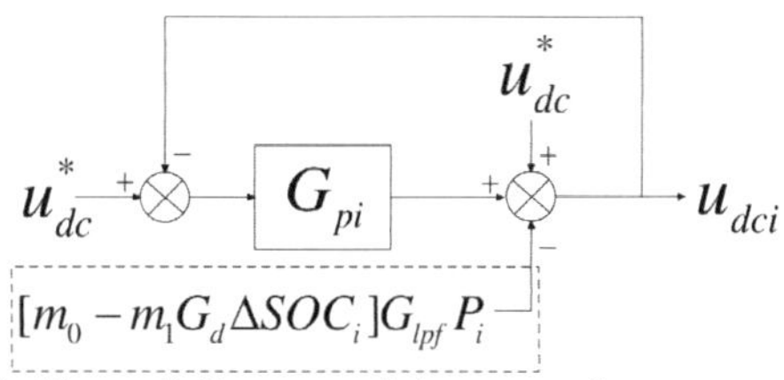

Figure 4. Control diagram of the secondary controller

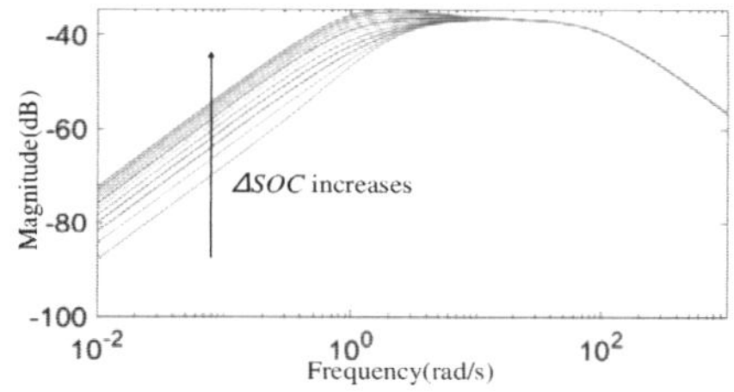

Figure 5. Frequency response of the secondary controller

3. Case Study and Simulation Test

In order to validate the feasibility of the proposed method, we use MATLAB/Simulink to set up the simulation model. Super capacitor is used as energy storage module and the rated capacity is 100F and rated voltage is 60V. The load power is 800W and DC bus rating is 200V. The SOC balance controller parameters m_0 is 0.0375, m_{i0} is 0.03. The PID parameters K_p and K_i are 1 and 5, respectively.

Firstly, consider the case of two modules. The initial values of SOC_1 and SOC_2 are set to 100% and 99%, respectively. As shown in the Figure 6 and Figure 7, P_1 is larger than P_2 under the control of the proposed method, which makes SOC_1 decreases faster than SOC_2. As the difference between SOC_1 and SOC_2 diminishes, P_1 and P_2 tends to be equal. And eventually, SOC balance is reached. It can also be seen from Figure 8 that the DC bus voltage can maintain the rating.

Also, the case of three modules is considered. The initial values of SOC_1, SOC_2 and SOC_3 are 100%, 99.3% and 99%, respectively. As the Figure 9 shown, before module 2 is removed, SOC_1, SOC_2 and SOC_3 have different falling speeds to reach the average. After module 2 is removed, which is a normal operation in practical applications, SOC_{avg} changes immediately since module 2 cannot be taken into account in the calculation of the average value anymore. SOC_2 and SOC_3 keep reaching balance gradually which means the proposed method still works after the disturbance. The stability of the proposed method is verified.

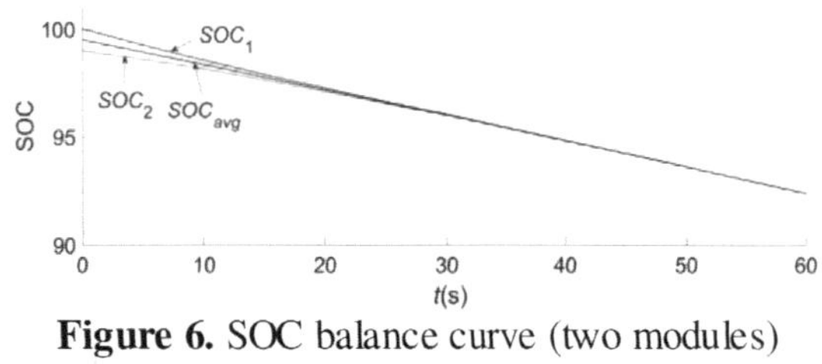

Figure 6. SOC balance curve (two modules)

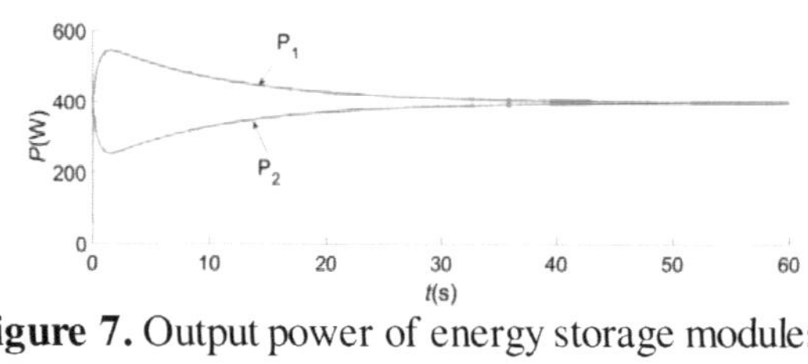

Figure 7. Output power of energy storage modules

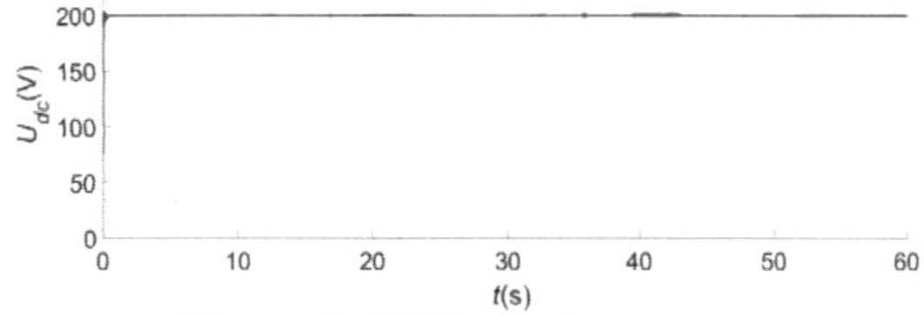

Figure 8. DC bus voltage curve

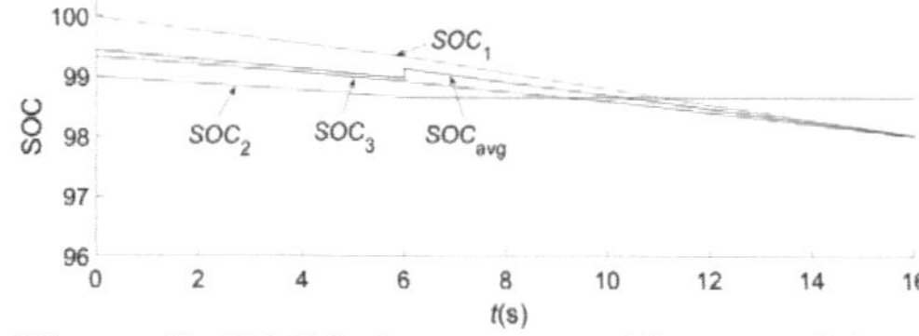

Figure 9. SOC balance curve (three modules)

4. Conclusions

In this paper, an improved droop control method is proposed to achieve SOC balance of different energy storage modules. The improved method introduces the low bandwidth network to measure the average SOC, which is used to regulate the droop coefficients. Under the control of the SOC balance controller, the SOC of different modules tend to be balanced and the output power will also be equal. The secondary voltage controller is also introduced to ensure the DC bus voltage maintains rating. MATLAB/Simulink simulation verifies the effectiveness of the proposed method.

5. References

[1] Wang P, Lu X, Yang X, Wang W and Xu D 2016 An Improved Distributed Secondary Control Method for DC Microgrids With Enhanced Dynamic Current Sharing Performance *IEEE Transactions on Power Electronics* **31** 6658-73

[2] Kakigano H, Nishino A, Miura Y and Ise T 2010 Distribution voltage control for DC microgrid by converters of energy storages considering the stored energy. In: *2010 IEEE Energy Conversion Congress and Exposition,* pp 2851-6

[3] She X, Huang A Q, Lukic S and Baran M E 2012 On Integration of Solid-State Transformer With Zonal DC Microgrid *IEEE Transactions on Smart Grid* **3** 975-85

[4] Dragičević T, Lu X, Vasquez J C and Guerrero J M 2016 DC Microgrids—Part II: A Review of Power Architectures, Applications, and Standardization Issues *IEEE Transactions on Power Electronics* **31** 3528-49

[5] Chunjiang Z, Jie D, Jun L and Bing B 2014 A Control Strategy for Battery-Ultracapacitor Hybrid Energy Storage System *Transactions of China Electrotechnical Society* **29** 334-40

[6] Xialin L, Li G, Chengshan W and Yunwei L 2016 Key Technologies of DC Microgrids:An Overview *Proceedings of the CSEE* **36** 2-17

[7] Shanshan Z, Fei W, Hui G, Qifeng W and Yanxia G 2018 Overview of Droop Control in DC Microgrid *Proceedings of the CSEE* **38** 72-84+344

[8] Dragičević T, Guerrero J M, Vasquez J C and Škrlec D 2014 Supervisory Control of an Adaptive-Droop Regulated DC Microgrid With Battery Management Capability *IEEE Transactions on Power Electronics* **29** 695-706

[9] Po-Hsu H, Weidong X and Moursi M S E 2013 A practical load sharing control strategy for DC microgrids and DC supplied houses. In: *IECON 2013 - 39th Annual Conference of the IEEE Industrial Electronics Society,* pp 7124-8

[10] Oliveira T R, Silva W W A G and Donoso-Garcia P F 2017 Distributed Secondary Level Control for Energy Storage Management in DC Microgrids *IEEE Transactions on Smart Grid* **8** 2597-607

[11] Ziwen L, Shihong M, Zhihua F, Kaiyun C and Yilong K Accurate Power Allocation and Zero Steady-State Error Voltage Control of the Islanding DC Microgird Based on Adaptive Droop Characteristics *Transactions of China Electrotechnical Society* 1-12

[12] Xiaonan L, Kai S, Lipei H, Xi X and M.Guerrero J 2013 Dynamic Load Power Sharing Method With Elimination of Bus Voltage Deviation for Energy Storage Systems in DC Micro-grids *Proceedings of the CSEE* **33** 37-46+20

Acknowledgments

This work was financially supported by Technical Standard Special Item of Shanghai Science and Technology Commission (18DZ2205700), Academician Expert Workstation Platform Project of Jinshan District (JSZJZ2018025G) and Industry-University-Research Cooperation Program of Minhang District, Shanghai (2017MH271).

Self-adaption Dynamic Virtual Inertia Control Strategy of Doubly fed Variable Speed Pumped Storage Unit

Zhichao Liu, Changhong Deng* and Yahong Chen

School of Electrical Engineering and Automation, Wuhan University

dengch@whu.edu.cn

Abstract. The power decoupling control characteristic of doubly fed variable speed pumped storage unit (VSPS) makes the rotor of VSPS unable to respond to the change of grid frequency and reduces the inertia of power system. In this paper, a self-adaption dynamic virtual inertia control strategy for VSPS is proposed. The strategy introduces correction factor to change the coefficient of control by detecting the change of $\mathrm{d}\Delta f/\mathrm{d}t$ and automatically control the active power output of VSPS to provide dynamic frequency support for power grid. Simulation results demonstrate that by dynamically compensating the reference power of doubly fed converter, the strategy can effectively increase the frequency stability of power grid.

1. Introduction

The intermittent and fluctuating characteristics of renewable energy would reduce the frequency stability of power system. The doubly fed induction generator (DFIG) based variable speed pumped storage unit (VSPS) has the advantages of flexible control and rapid power response [1], and becomes one of the most economical and effective means to ensure the safe and stable operation of power system.

However, the power decoupling control characteristic of doubly fed converter makes the rotor of DFIG in VSPS unable to respond to the grid frequency change thus VSPS contributes no inertia to the whole inertia of power system [2]. The direct consequence of this is that the frequency stability of the system decreases. In order to improve overall inertia and therefore improve frequency stability of power system, applying virtual inertia control to VSPS is therefore needed [3].

In the existing literatures, proportional differential (PD) controller is most commonly adopted by the virtual inertia control loop of doubly fed converter for DFIG based wind turbine generator. In Reference [4] control loop of $\mathrm{d}f/\mathrm{d}t$ is replaced by $\mathrm{d}\Delta f/\mathrm{d}t$ to compensate for the negative damping of VSPS that is introduced by traditional inertial control. In Reference [5] the variable droop coefficient strategy is used to change the distribution of active power output of generation units in the system, which successfully improves the lowest frequency of the system.

However, despite the fact that virtual inertia control of DFIG based variable wind turbine generator have been extensively investigated in vast literatures, there is a serious shortage of research in the area of virtual inertia control of VSPS. Besides, the existing control strategy of VSPS are generally adopt fixed control coefficients, the advantages of large adjustable range of unit speed are neglected [6], and the potential of enhancing frequency stability of power system by fully tapping the real-time available capacity of unit is neglected.

In this paper, a self-adaption dynamic (SAD) virtual inertia control strategy for VSPS is proposed. The strategy introduces correction factor to change the coefficient of control by detecting dΔf/dt and automatically control the active output of VSPS to provide dynamic frequency support for power grid. Simulation results demonstrate that compared with droop control and fixed coefficient virtual inertia control, VSPS with SAD virtual inertia control has better frequency response capability and stronger inertial support capability for power system, which verifies the feasibility of the proposed strategy.

2. Modeling of VSPS

2.1. Overall structure of VSPS

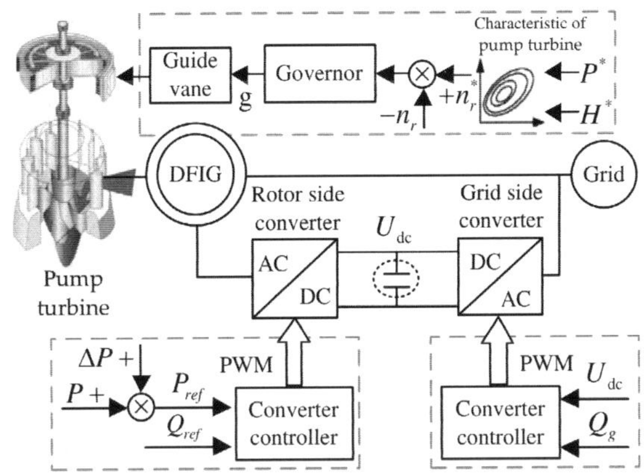

Figure 1. Structure of VSPS.

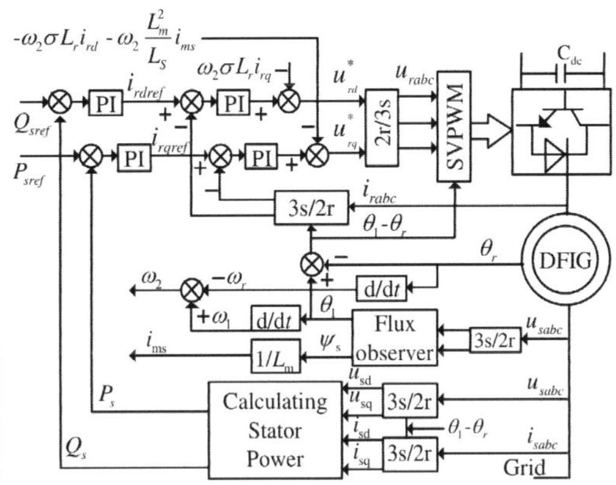

Figure 2. Diagram of rotor side control.

VSPS is a new type of hydraulic power generation system, which consists of reversible pump turbine unit, speed governor, doubly fed converter and so on. The stator of DFIG is connected to grid, and the three-phase symmetrically distributed excitation windings are used on rotor where variable frequency power supply device provides symmetrical AC excitation to it. The magnitude, frequency, phase and phase sequence of excitation voltage can be controlled according to the requirements, so that the magnitude and position of excitation magnetic field relative to rotor and the speed of DFIG can be controlled. Strong controllability makes DFIG have high operation stability, variable speed and constant frequency power generation capacity and independent active and reactive power regulation capacity [7]. The system diagram of VSPS is shown in figure 1. The optimal speed reference of VSPS is calculated according to the nonlinear characteristic of pump turbine and reference power P^* of VSPS and net head H^* of water reservoir. The speed control of VSPS is realized by adjusting the guide vane opening through speed governor. Stator flux orientation control is adopted by double fed converter to control the electromagnetic power of DFIG according to the power instruction. In the diagram, P_{ref} is rotor side active power reference command and ΔP is additional power command by control strategy.

2.2. Model for rotor side control of DFIG converter

On the premise of applying stator flux orientation control for DFIG and decoupling control for double fed converter, the active and reactive power of DFIG are regulated by stator current i_{ds} and i_{qs} respectively. For the sake that i_{ds} and i_{qs} are directly controlled by the rotor side converter of doubly fed converter, and the main control purpose of the grid side converter is to maintain the DC-side capacitor voltage stability, only the control block of rotor side is given, as is shown in figure 2. In the diagram of rotor side control, double closed-loop control structure is adopted where the outer loop is power control loop, which is mainly responsible for controlling the active and reactive power of DFIG, and the inner loop is current control loop, which is mainly responsible for controlling the d and q axis currents for DFIG, respectively.

3. Virtual inertia control of VSPS

3.1. Virtual inertia control of VSPS

By adding ΔP to the reference value of active power for the rotor side converter of VSPS and releasing the kinetic energy stored in the rotor to actively respond to the frequency change of power system, the frequency response characteristic of conventional generator can be simulated and the inertia of the system can be increased. Power of VSPS in transient state is affected by coefficient of differential link. The additional active power ΔP of virtual inertia control is expressed as follows:

$$\Delta P = (K_p \Delta f + K_d \frac{d\Delta f}{dt}) \tag{1}$$

Where Δf is system frequency deviation, and K_p and K_d are coefficients of proportional and differential control, respectively.

3.2 Relationship between virtual inertia control and frequency stability of power system

Frequency stability of power system is directly affected by system inertia and damping [3]. The greater the inertia is, the lower the frequency offset and the initial frequency change rate are, but the adjustment time of frequency stabilization increases. With the increase of system damping, the frequency offset decreases, but the speed of frequency reaching peak value accelerates and the initial frequency change rate increases. The rotor motion equation of VSPS is as follows:

$$(2H_\mathrm{S} + K_\mathrm{d})\frac{\mathrm{d}\Delta f}{\mathrm{d}t} = P_\mathrm{m} - P_\mathrm{e} - (D + K_\mathrm{p})\Delta f \tag{2}$$

Where H_s is inherent inertia constant of VSPS, D is damping coefficient, P_m is mechanical power of prime mover and P_e is electromagnetic power of VSPS.

K_p and K_d directly determine the damping and inertia of virtual inertia control introduced into the system by VSPS, and jointly determine the frequency response characteristics of the system [8]. K_p produce system damping, while K_d produce a similar rotational inertia as synchronous generator. With the increase of K_p, the frequency nadir (FN) and recovery speed of the system increase, the short-term power of VSPS increases, and the steady-state rotation speed of VSPS decreases. K_d determines the inertial response characteristics of the system. The larger the K_d, the slower the system frequency descends, the longer the time to reach FN and the oscillation time, the higher the short-term power of VSPS, while the steady-state rotation speed of VSPS remains unchanged.

4. Self-adaption dynamic (SAD) virtual inertia control

When load disturbance occurs in the system, the frequency will change from the original stable value to the new stable value, and the frequency oscillation will occur at the moment of the disturbance. The typical frequency change curve can be divided into three stages. The characteristics of frequency variation and the required virtual inertia are different in each stage.

In stage 1, the load suddenly increases, the frequency drops rapidly to reach FN, $\mathrm{d}\Delta f/\mathrm{d}t$ suddenly increases from 0 to the maximum value, and falls back in a very short time, and decreases to 0 in FN. At this stage, $\Delta f<0$ and $\mathrm{d}\Delta f/\mathrm{d}t<0$, it is necessary to increase the inertia of VSPS and power system by increasing the power reference value of rotor side converter of VSPS, so that the rotor can release kinetic energy quickly, restrain the speed of the system frequency drop and improve FN.

In stage 2, Δf begins to decrease slowly, and $\mathrm{d}\Delta f/\mathrm{d}t$ crosses zero first and then reverses. At this stage, $\Delta f<0$ and $\mathrm{d}\Delta f/\mathrm{d}t>0$, at the same time, the rotor of VSPS enters the acceleration stage. In this stage the value of virtual inertia should be appropriately reduced to speed up frequency recovery.

In stage 3, the frequency tends to be stable and the value of virtual inertia should be stable.

To meet the requirements, the additional power provided by the virtual inertia control of VSPS, i.e. the virtual inertia of VSPS, is automatically and dynamically adjusted by the correction factor α. The relationship between α and Δf and $\mathrm{d}\Delta f/\mathrm{d}t$ is as follows.

$$\begin{cases} \alpha = 1, & |\frac{d\Delta f}{dt}| < K_1 \\ \alpha = 1 + k\frac{d\Delta f}{dt}, K_1 \le |\frac{d\Delta f}{dt}| \le K_2 \\ \alpha = 1 + kK_2, & |\frac{d\Delta f}{dt}| > K_2 \end{cases} \tag{3}$$

Where k is the proportionality coefficient, K_1 and K_2 are the interval separating values of $d\Delta f/dt$. K_1 can avoid the coefficient oscillation caused by measurement error and ensure the stability of the system in steady state. K_2 can limit the over-response of unit caused by excessive df/dt, thus avoiding unit power oscillation.

After introducing the correction factor, the additional active power reference value ΔP provided by SAD virtual inertia control is as follows.

$$\Delta P = \alpha K_p \Delta f + K_d \frac{d\Delta f}{dt} = K_p \Delta f + K_d \frac{d\Delta f}{dt} + k\Delta f \frac{d\Delta f}{dt} = \Delta P_p + \Delta P_d + \Delta P_{ad} \tag{4}$$

In addition to the proportional part ΔP_p and differential part ΔP_d provided by traditional virtual inertia control, the additional power ΔP_{ad} is added in ΔP by SAD virtual inertia control. In the initial phase of frequency drop, the sign of Δf and $d\Delta f/dt$ are identical, ΔP_{ad} is positive, and the power reference value of VSPS increases correspondingly. The SAD control has a larger electromagnetic power increment than traditional virtual inertia control of the same control parameter, which can more effectively suppress the frequency drop and provide greater inertial support. On the contrary, in the stage where the frequency starts to recover, the sign of Δf and $d\Delta f/dt$ are different, therefore ΔP_{ad} is negative. The power reference value is consequently becomes smaller, and the active power output of VSPS is reduced appropriately, which avoids the severe power oscillation of the unit.

5. Simulation analysis

To verify the effectiveness of SAD virtual inertia control, an isolated power system including 100 MW wind farm G_1, 10 MW diesel engine G_2, two 100 MW VSPSs G_3 and G_4, and loads L_1 and L_2 is built. Under initial condition, load is 100 MW, wind speed keeps 10 m/s unchanged, and the system operates in steady state. At 1s, 20 MW impulse load is put into operation. Under the same simulation conditions, VSPS is controlled by traditional droop control, traditional virtual inertia control and SAD virtual inertia control, and the results are compared.

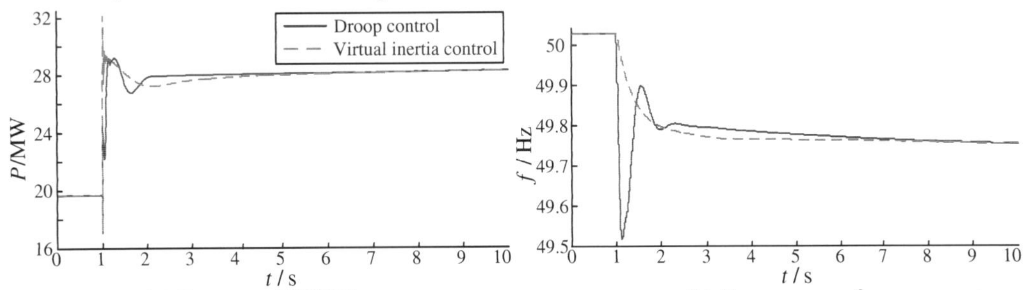

(a) Power of VSPS. (b) Frequency of power system.

Figure 3. Frequency and power response of VSPS.

5.1. Comparison of different inertia control strategies

The frequency and power response curve of VSPS is shown in figure 3. The coefficient of droop is 1/0.05, K_p =1/0.05, and K_d =15. When applying traditional droop control, FN is 49.52Hz, the maximum output of the unit is 29.2MW, and the adjustment time is 7s. When applying traditional virtual inertia control, FN is 49.75Hz, the maximum output of the unit is 44.4MW, and the adjustment time is 4s. Compared with traditional droop control, FN of traditional virtual inertia control increased by 0.25Hz and the adjustment time is 42.8% faster. It is obvious that, by using virtual inertia control strategy, falling speed of the system frequency slows down and the frequency oscillation is restrained.

The frequency and power response curves of fixed coefficient virtual inertia control and SAD virtual inertia control are shown in figure 4. In all cases, K_p = 1/0.05, however, K_{d1} = 15 for small inertia control, K_{d2} = 30 for large inertia control, and K_{d3} = 15 for SAD control.

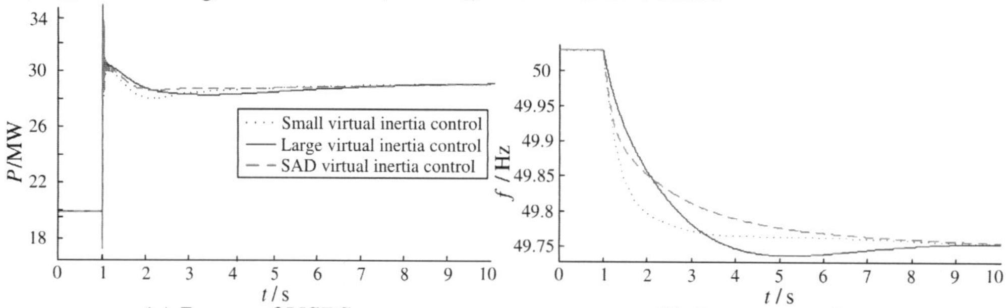

(a) Power of VSPS. (b) Frequency of power system.

Figure 4. Frequency and power response of virtual inertial control.

In the phase of frequency drop, the maximum power output of VSPS in the case of small inertia control is 29.5 MW, in the case of large inertia control is 34.5 MW, and in the case of SAD control is 34 MW. Compared with the small inertia control with the same control coefficient, the maximum instantaneous power output of VSPS by SAD control increased by 15.25%. The effect of suppressing frequency drop by SAD control is more obvious, and the effect of inertia response is close to that of large inertia control. In the phase of frequency recovery from FN to stable value, FN of large inertia control is 49.73Hz, and adjusting time is 7s; FN of SAD control increased to 49.76Hz, and adjusting time is 5s. By dynamically reducing the inertia of VSPS, SAD control avoids the oscillation caused by over-response of the unit, and the frequency adjustment time becomes shorter, which speeds up the frequency recovery of the system.

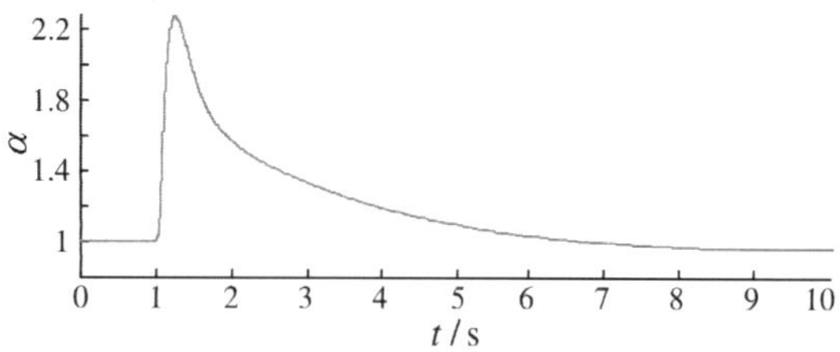

Figure 5. Correction factor by SAD control.

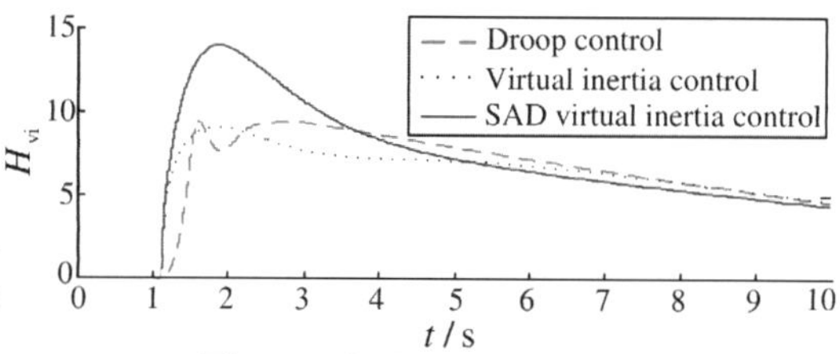

Figure 6. Comparison of H_{vi}.

Correction factor α by SAD control is shown in figure 5. α undergoes a dynamic process from rapid increase to gradual decay to steady state. During the phase of frequency drop, $d\Delta f/dt$ increases rapidly and SAD virtual inertia control responds quickly correspondingly. The power reference of VSPS is adjusted by increasing α, and then the virtual inertia of the system is increased. After the frequency drops to FN, it enters the frequency stabilization stage, $d\Delta f/dt$ is quite small. The frequency is stabilized quickly by reducing the equivalent inertia through decreasing α. In the stable stage, $d\Delta f/dt$ approaches 0, α approaches 1, and the response process of SAD virtual inertia control ends.

5.2. Comparison of equivalent virtual inertia time constants

The equivalent inertia time constant H_{vi} of VSPS with different frequency regulation strategies is shown in figure 6. The value of H_{vi} is calculated according to $\Delta\omega_r/\Delta\omega_s$ in the process of inertia response by VSPS [8]. It can be seen that H_{vi} experienced a dynamic process from initial oscillation to gradual attenuation to steady-state value, with a three-stage variation characteristic.

In the phase of fast response: the frequency begins to fall. The inertia control makes the speed of unit decrease rapidly and releases the kinetic energy. The rotor speed decreases faster than the system frequency decreases, which increases $\Delta\omega_r/\Delta\omega_s$ and increases H_{vi}. Compared with traditional droop control and traditional inertia control, SAD virtual inertia control has the fastest response speed, and H_{vi} quickly reaches the maximum value.

In the phase of speed recovery: the speed starts to rise, which makes $\Delta\omega_r/\Delta\omega_s$ decreases and H_{vi} begins to decrease. Compared with traditional droop control and traditional inertia control, by SAD virtual inertia control, the absolute value of speed change is larger, H_{vi} drops more obviously. In additional to that, H_{vi} has obvious oscillation process in traditional droop control.

In the phase of stable: as the system frequency tends to stabilize and the speed of VSPS returns to the rated value, H_{vi} gradually approaches the steady state value, and the process of inertial response gradually ends.

6. Conclusion

In order to achieve better dynamic performance of frequency response for power system, a self-adaption dynamic virtual inertia control strategy for VSPS is proposed. A correction factor is introduced to change the coefficient of control and automatically control the active power output of VSPS to provide dynamic frequency support for power grid. By dynamically changing the virtual inertia at different stages the frequency response characteristics of VSPS can be better improved. By using this method, VSPS has better inertia support ability to the power grid, and can effectively enhance the frequency stability of power system.

7. References

[1] M. Valavi and A. Nysveen, "Variable-Speed Operation of Hydropower Plants: A Look at the Past, Present, and Future," in IEEE Industry Applications Magazine, vol. 24, no. 5, pp. 18-27, Sept.-Oct. 2018.

[2] M. A. Bidgoli, H. A. Mohammadpour and S. M. T. Bathaee, "Advanced Vector Control Design for DFIM-Based Hydropower Storage for Fault Ride-Through Enhancement," in IEEE Transactions on Energy Conversion, vol. 30, no. 4, pp. 1449-1459, Dec. 2015.

[3] M. F. M. Arani and E. F. El-Saadany, "Implementing Virtual Inertia in DFIG-Based Wind Power Generation," in IEEE Transactions on Power Systems, vol. 28, no. 2, pp. 1373-1384, May 2013.

[4] Hui LI, Liu H, et al. Improved Virtual Inertia Control Strategy of Doubly Fed Pumped Storage Unit for Power Network Frequency Modulation[J]. Automation of Electric Power Systems, 2017, 41(10):58-65.

[5] Hui LI , Kun W , Haitao L , et al. Variable droop coefficient frequency control strategy of AC excited pumped storage unit[J]. Electric Power Automation Equipment, 2018.

[6] A. Joseph and T. R. Chelliah, "A Review of Power Electronic Converters for Variable Speed Pumped Storage Plants: Configurations, Operational Challenges, and Future Scopes," in IEEE Journal of Emerging and Selected Topics in Power Electronics, vol. 6, no. 1, pp. 103-119, March 2018.

[7] T. Kuwabara, A. Shibuya, H. Furuta, E. Kita and K. Mitsuhashi, "Design and dynamic response characteristics of 400 MW adjustable speed pumped storage unit for Ohkawachi Power Station," in IEEE Transactions on Energy Conversion, vol. 11, no. 2, pp. 376-384, June 1996.

[8] MIGUELA, TORRESL, LOPESLAC, etal. Self-tuning virtual synchronous machine: a control strategy for energys to rage systems to support dynamic frequency control [J]. IEEE Transactions on Energy Conversion, 2014, 29 (4): 833-840.

Acknowledgments

This work was supported by The National Key Research and Development Program of China (2017YFB0903705).

IOP Conf. Series: Earth and Environmental Science **342** (2019) 012010 doi:10.1088/1755-1315/342/1/012010

Determination of maximum level of EV penetration with consideration of EV charging load and harmonic currents

Chengke Zhou[1], Hang Wang[2], Wenjun Zhou[2], Kejun Qian[3], Shaoxin Meng[2,4]

[1]School of computing, engineering and built environment, Glasgow Caledonian University, Glasgow, UK
[2]College of electrical engineering and automation, Wuhan University, 8th South Road of East Lake, Wuhan, China
[3]Suzhou power supply company, state grid corporation of China, Suzhou, China
[4]State key laboratory of power grid environmental protection, Wuhan, China

Email: whuwanghang@qq.com

Abstract. In order to accommodate Electric Vehicles (EVs) charging demand which is expected to rise steeply in the coming years due to environmental concerns on global warming, it is important to determine the maximum EV penetration can be allowed without network upgrading and avoid overloading of Medium Voltage (MV) cables and distribution transformers in residential distribution power systems. Previous work on modelling and analysis of relationship between the temperature of MV cables and transformer with EV penetration level failed to consider the contribution of harmonic distortion in the demand. In this paper, a methodology is proposed to model the EV penetration and their effects on operational temperature of MV cables and transformers, with charging scenarios and harmonic currents taken into consideration. The method is simulated in a typical urban residential distribution system. The results show that EV penetration above 27.25 % with 6.6 kW EV chargers charging in uncontrolled domestic scenario will lead to 10 kV 400 mm^2 three core cable overloading in the simulation with harmonic current taken into consideration. The result is 30.74 % if harmonic currents are ignored. The maximum allowable EV penetration of 35/10 kV 16 MVA and 10 kV 1250 kVA transformer is 50.22% and 26.19% respectively. If EVs charge in off-peak time, the cable and transformer can avoid overloading and the charging cost is lower.

1. Introduction

The rapid growth of EV charging demand increases the overload risk of equipment in distribution networks (DNs) which are already heavily loaded in many urban residential areas in countries around the world. Especially in China where most metropolises are of temperate or subtropical monsoon climate, and the peak demand and ambient temperature appear simultaneously in summer because of the high usage of air conditioners. The load demand, temperature increase and thermal aging of cables and transformers caused by EV charging have been modelled by the present authors in literatures [1-3]. However, their earlier work and reported by others [4-5] may have underestimated the impact of EV charging, because of the ignorance of additional AC loss due to the harmonic current caused by AC/DC converters. Consequently, without consideration of harmonic current due to EV charging may lead to cable and transformer overloading. Furthermore, the rated ampacity of cables and capacity of transformer have been determined by traditional model with consideration of future power load increases when a DN was designed [6]. On account of 20-40 years designed life, it is essential to allow derating of MV cables and transformers ampacity caused by the fast emerging non-sinusoidal loads such as EV chargers to be modelled to avoid overloading. This paper aims to model maximum level of EV penetration allowable in a distribution system and analyse its impact on operational temperature of MV cables and transformers with harmonic current taken into consideration.

2. Mathematic models

The capacity of electrical devices such as cables and transformers in distribution system are influenced by the level of load as well as harmonic components. With the increase in harmonic orders, active power loss increases because of skin effect and proximity effect. In order to investigate the impacts on cables and transformers, the load power with harmonic component is defined in Equation 1, in which P_h and P_l is harmonic power per unit, fundamental power per unit, Ii is current of i th order, n is the maximum harmonic order .

$$\frac{P_h}{P_s} = \frac{\sqrt{\sum_{i=1}^{n} I_i^2}}{I_1} \quad (1)$$

Then, the equations of cables and transformers temperature are deduced from the equations provided by IEC 60287 and IEEE Std C57.110 [7, 8]. Because the magnitude of harmonic currents above 19th due to EV charging is low, their thermal effect on MV cables and transformers is ignored in this paper. The temperature model of a three core cable carrying non-sinusoidal current is presented in Equation 2, in which θ_c is conductor temperature of cable, R_{AC} is the AC resistance of conductor, c_n is the number of conductors and λ_2 is loss ratio of armour layer. If the thermal resistances T_1, T_2, T_3, T_4 and ambient temperature θ_A become available via either measurement or calculation, the conductor temperature in Equation 2 can be obtained.

$$\theta_c = \sum_{1}^{n} \frac{I_i^2}{I_1^2} \times R_{AC}(i)\left(T_1 + c_n T_2 + c_n\left(1+\lambda_2(i)\right)\left(T_3+T_4\right)\right) + \theta_A \quad (2)$$

At any time instant t, the hot-spot temperature θ_g of transformer consists of three components defined in Equation 3, where θ_A, $\Delta\theta_{TO}$ and $\Delta\theta_g$ are the ambient temperature, then top-oil temperature rise over ambient temperature, and the winding hotspot rise over top-oil temperature, respectively, during the time interval Δt in degree Celsius.

$$\theta_g = \theta_A + \Delta\theta_{TO} + \Delta\theta_g \quad (3)$$

The top oil temperature rise above ambient temperature $\Delta\theta_{TO}$ with sinusoidal load is defined in Equation 4, in which $\theta_{TO\text{-}R}$ is top oil temperature rise at rated load, Δt is time interval, R is the ratio of load loss to no-load loss, $\tau_{TO\text{-}R}$ is the oil time constant at rated condition in hours, p is an empirically derived exponent to account for effects of change in resistance with change in load.

$$\frac{\Delta\theta_{TO}(t)}{\theta_{TO\text{-}R}} = \left[\left(\frac{K^2(t+\Delta t)R+1}{R+1}\right)^p - \left(\frac{K^2(t)R+1}{R+1}\right)^p\right] \times \left(1 - exp\left(\frac{-\Delta t\left[\left(\frac{K^2(t+\Delta t)R+1}{R+1}\right)^{\frac{1}{p}} - \left(\frac{K^2(t)R+1}{R+1}\right)^{\frac{1}{p}}\right]}{\tau_{TO.R}\left[\frac{K^2(t+\Delta t)R+1}{R+1} - \frac{K^2(t)R+1}{R+1}\right]}\right)\right) + \left(\frac{K^2(t)R+1}{R+1}\right)^p \quad (4)$$

Similarly, the hot-spot temperature rise to top oil temperature is defined in Equation 5, in which $\theta_{g\text{-}R}$ is the hot-spot temperature rise over top-oil temperature at rated load, τ_m is the winding time constant in hours, m is an empirically derived exponent to account for effects of change in resistance and oil viscosity with change in load.

$$\frac{\Delta\theta_g(t)}{\theta_{g\text{-}R}} = \left(K^{2m}(t+\Delta t) - K^{2m}(t)\right) \times \left(1 - exp\left(-\frac{\Delta t}{\tau_w}\right)\right) + K^{2m}(t) \quad (5)$$

The model of transformer is formulated by power loss per unit. The power loss of transformer are defined by Equation 6, in which P_{LL} is the load loss, P is conductor loss portion of the load loss, F_{HL} is harmonic loss factor for winding eddy current, P_{EC} is the winding eddy-current loss, $F_{HL\text{-}STR}$ is the harmonic loss factor for other stray losses, P_{OSL} is the other stray loss.

$$P_{LL} = P + F_{HL} \times P_{EC} + F_{HL\text{-}STR} \times P_{OSL} \tag{6}$$

The method to calculate F_{HL} and $F_{HL\text{-}STR}$ is provided by IEEE std 57.110. If the magnitude and order of harmonic current can be obtained via measurement or simulation, the P_{LL} can be calculated. The load ratio $K(t, n)$ at time instant t in presence of harmonic current is defined in Equation 7. Using the $K(t, n)$ to change the ratio of the load K (t) at time t, $\Delta\theta_{TO}$ and $\Delta\theta_g$ in presence of harmonic load can be obtained.

$$K(t,n) = \frac{P_{LL}(t,n)}{P_{LL-R}} (0 \le t \le 24) \tag{7}$$

The maximum level of EV penetration allowable in residential distribution network is determined by cable conductor temperature, hot-spot temperature, hot-spot temperature rise and top oil temperature of transformers.

3. Simulations

A generic urban distribution network which is a representative of urban residential distribution network in China is presented in Fig.1. The system consists of a 500 MVA 35 kV three phase power source, a 10 kV substation, a 35/10 kV 16 MVA transformer (T1) and seven 10/0.4 kV YY0 1250 kVA transformers(T2) and three 10 kV outgoing feeders (F_1, F_2 and F_3). Three phase of a 10 kV cable is consisted of F_1, F_2 and F_3. The power frequency is 50 Hz and the power factor of residential load is set at 0.9.The load ratio of residential load to the rated load of 10 kV 400 mm^2 is 0.7, the maximum phase current r.m.s value of 10 kV cable in presence of residential load is 372.24 A. The EV charging load changes with EV penetration. We assume that the three phase cables and transformers are balanced. The total number of vehicles in the studied region is 1246 cars and the EV penetration is set vary from 0 to 70 %. A program is developed in MATLAB based on the methodology presented in this paper. The following data are used in the simulation. The EV can be fully charged in 4 hours at most with a charging power 6.6 kW. The hourly ambient temperature and residential load ratio of a typical summer day is given in Table A1 in the appendices of the paper. Harmonic current of a single phase rectifier and thermal characteristics of cable and transformer are presented in Table A2 and A3 in appendices. In Wuhan, China, most workers start work at 9:00 and go home at 17:00. It is assumed that all the EV users go home after work in this paper, to allow the worst scenario (scenario 1: uncontrolled domestic charging) to be considered. Furthermore, accounting for the average commuting time of EV in Wuhan of 47 minutes, all the EV users are expected to begin their recharge cycles every work day at nearly the same time (at 18:00). The time-of-use electricity rate structure is a typical tariff adopted by electric power company to encourage EV users to charge during the off-peak time. In uncontrolled domestic off-peak charging scenario (scenario 2), it is assumed that all the EV users charge during off-peak time. For instance, in Wuhan, the peak load time is 7:00 to 21:59 while off-peak load time is 22:00 to 6:59 in the next day. In this scenario, the EV owners would charge during the off-peak time and the battery of EVs should be fully charged at the end of off-peak load time.

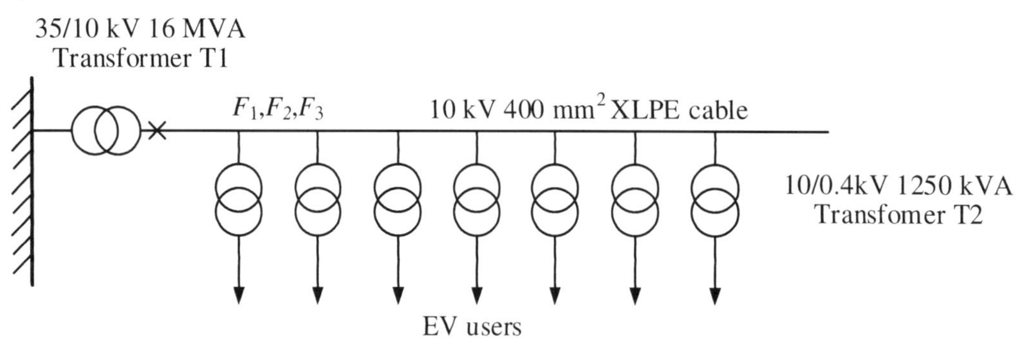

Fig.1 Residential MV distribution system used in the simulation

Using Equation 1-7 with values taken from Table A1-A3 the results of allowable maximum EV penetration shown in Table 1 in the simulated system can be obtained. The results in Table 1 show that, with the given condition of cable conductor temperature, the maximum EV penetration decreases by 3.49% if harmonic currents due to EVs charging are taken into consideration. The deciding factor of the transformers are hot-spot temperature in the simulation. Due to limitation in the hot-spot temperature of T1 and T2, the maximum EV penetration will decrease by 5.48% and 2.90% respectively if harmonic currents are taken into consideration.

Table 1 Allowable maximum EV penetration in simulation confined by permissible temperature

	Cable		T1		T2	
With harmonic current	Yes	No	Yes	No	Yes	No
θ_c	27.25 %	30.74 %	–	–	–	–
θ_g	–	–	50.22%	55.70%	26.19 %	29.05 %
$\Delta\theta_{TO}$	–	–	51.46%	57.09%	27.64 %	30.65%
$\Delta\theta_g$	–	–	55.42%	61.48%	30.32 %	33.64%
All conditions	27.25 %	30.74 %	50.22%	55.70%	26.19 %	29.05%

The results of hourly temperature in typical scenarios are presented in Fig.2. In uncontrolled domestic charging scenario, the results in Fig.2 (a1-a4) show that, the peak of the conductor temperature of cable, θ_g, $\Delta\theta_{TO}$, $\Delta\theta_g$ appears around 19:00. During the charging time commencing from 18:00 to 22:00, the aforementioned temperatures increase significantly. In off-peak uncontrolled domestic charging scenario, the results in Fig.2 (b1-b4) show that, the peak of the conductor temperature of cable, θ_g, $\Delta\theta_{TO}$, $\Delta\theta_g$ appears also around 19:00, whilst the charging load peak is around 22:00 to 2:00 of the next day. By analysing the results of Fig.2, it can be concluded that cables and transformers can avoid overloading under 30% EV penetration in off-peak uncontrolled domestic charging scenario while the charging cost is lower than uncontrolled domestic charging scenario because of time of use electric price.

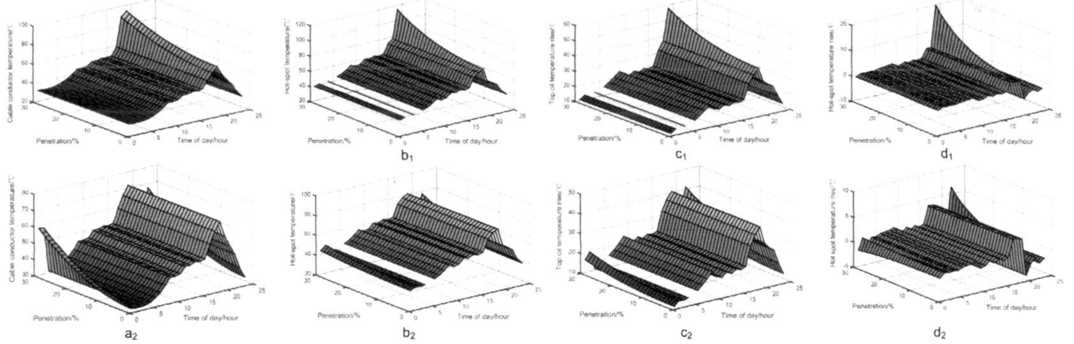

Fig.2 Cable and Transformer temperature in summer in typical charging scenarios (Scenario x(x=1 or 2): a_x cable conductor temperature; b_x, hot-spot temperature of 10/0.4 kV transformer; c_x, top oil temperature of 10/0.4 kV transformer; d_x, hot-spot temperature rise of 10/0.4 kV transformer)

4. Conclusions

An analytical model is presented to evaluate maximum EV penetration level in a residential distribution system and EV charging impacts on cable and transformer temperature with harmonic current taken into consideration in this paper. Through simulations and analysis, the paper comes into the following conclusions. The capacity of distribution system will decrease due to EVs charging which can be exacerbated when considering the presence of harmonic current. The results of the simulation show that 27.25 % penetration of 6.6 kW EV chargers charging in uncontrolled domestic charging scenario will lead to 10 kV 400 mm^2 three core cable overloading with harmonic current taken into consideration. The result is 30.74 % if harmonics are ignored. The maximum allowable EV penetration of 35/10 kV 16MVA and 10 kV 1250 kVA transformer is 50.22% and 26.19% respectively. If EVs charge in off-peak time, overloading of the cable and transformer can be avoided and the charging cost is lower. In the future work, the capacity of HV and LV power systems to adopt EV charging load in the presence of harmonic current will be modelled.

Acknowledgments
This work is supported by National Natural Science Foundation of China (51607098) the Science and Technology Project of State Grid Jiangsu Electric Power Company (J2018071) and the Open Fund of State Key Laboratory of Power Grid Environmental Protection (GYW51201801174), China.

References
[1] Qian K, Zhou C, Allan M and Yuan Y 2011 Modeling of load demand due to EV battery charging in distribution systems. *IEEE Trans. on Power Systems.*
[2] Zhang P, Qian K, Zhou C, Stewart B G and Hepburn D M 2012 A methodology for optimization of power systems demand due to electric vehicle charging load. *IEEE Trans on Power Systems.*
[3] Qian, K., Zhou, C., & Yuan, Y. (2015). Impacts of high penetration level of fully electric vehicles charging loads on the thermal ageing of power transformers. *J. Electr. Power & Energy Systems.*
[4] Razeghi G, Zhang L, Brown T, and Samuelsen S 2014 Impacts of plug-in hybrid electric vehicles on a residential transformer using stochastic and empirical analysis. *J. Power Sources.*
[5] Clement-Nyns K, Haesen E and Driesen J 2010 The impact of charging plug-in hybrid electric vehicles on a residential distribution grid. *IEEE Trans. on Power Systems.*
[6] IEEE C57.91-1995 IEEE guide for loading mineral-oil-immersed transformer.
[7] IEC 60287 Electric cables - Calculation of the current rating.
[8] IEEE C57.110-2008 IEEE recommended practice for establishing liquid-filled and dry-type power and distribution transformer capability when supplying non-sinusoidal load currents.

Appendices

Table A1 Hourly ambient temperature and residential load ratio in a typical summer day of Wuhan

	θ_A/ °C	load/p.u.		θ_A / °C	load/p.u.		θ_A / °C	load/p.u.		θ_A / °C	load/p.u.
1	29	0.26	**7**	29	0.475	**13**	38	0.57	**19**	37	1
2	28	0.30	**8**	30	0.55	**14**	38	0.60	**20**	35	0.88
3	28	0.30	**9**	31	0.63	**15**	38	0.59	**21**	33	0.80
4	28	0.32	**10**	33	0.58	**16**	38	0.61	**22**	30	0.75
5	29	0.32	**11**	35	0.55	**17**	37	0.80	**23**	28	0.60
6	29	0.40	**12**	37	0.59	**18**	37	0.92	**24**	28	0.42

Table A2 Harmonic current of a single phase rectifier EV charger

Harmonic order	Current magnitude /p.u.	Harmonic order	Current magnitude /p.u.
1	1.00	**11**	0.07
3	0.32	**13**	0.06
5	0.18	**15**	0.05
7	0.12	**17**	0.04
9	0.09	**19**	0.03

Table A3 Thermal characteristics of 10kV 400 mm^2 XLPE three core cable and transformers

	T1	T2		Cable
$\Delta\theta_{TO\text{-}R}$	65 ℃	55 ℃	θ_{max}	90 ℃
$\Delta\theta_{g\text{-}R}$	30 ℃	20.3 ℃	T_1	0.2129 K·m/W
$\theta_{T\text{-}R}$	125 ℃	105 ℃	T_2	0
p	0.9	0.9	T_3	0.1591 K·m/W
m	0.8	0.8	T_4	1.1417 K·m/W
$\tau_{TO\text{-}R}$	3.5 h	3.0 h	I_{rated}	531.77 A
τ_w	4.8 min	10 min		
R	5.96	8.76		

IOP Conf. Series: Earth and Environmental Science **342** (2019) 012011 doi:10.1088/1755-1315/342/1/012011

An improved particle swarm optimization algorithm and its application on distribution generation accessing to distribution network

Zheng Wei[1], Zhi Yong[1], Liang Chen[1], Gao Lei[1] and Zhu Wenpei[2]

[1] Institute of Electric Power Scientific Research, State Grid of Gansu Province, Lanzhou 730050, P.R. China
[2] Lanzhou University of Technology, Gansu Province, Lanzhou 730050, China

Email: 5693665@qq.com

Abstract. To seek the optimal way of distributed generation(DG) accessing to the distribution network(DN) is becoming a hot topic in the research of modern smart grid. Aims to the drawbacks of imbalance and the sluggish convergence speed of some traditional intelligent algorithms, such as, particle swarm optimization algorithm, when it is used to solve DG node optimal assignment, an improved particle swarm optimization algorithm with contraction factor based on the hybrid simulated annealing algorithm is proposed. Using the improved method, the minimum power loss can be guaranteed when the DGs are accessed to the DN. The influence of the DN on the grid voltage and the network loss of the distribution of the distributed generation are analyzed. The optimization model of the active power loss minimization is derived. The simulation results are used to verify that the improved algorithm is advantageous because of its a balanced global search ability as well as fast convergence speed to assign the different DG access nodes optimally.

1. Introduction

The demand for renewable electricity is becoming more and more fashionable to meet the rapid development of the current economy. In the face of increasing residential electricity consumption and high industrial energy consumption for production[1]. However, the environmental protection requirement is becoming more and more serious, for the traditional power generation, a single centralized mode of power supply is faced with many problems, depleted fossil energy is highly polluting and uneconomic[2]. It no longer meets the needs of the current rapid development. As the inexhaustible renewable wind and solar energies, these distributed electric sources not only can save cost and reduce energy consumption, but also can track load changes in time, improve system stability and meet environmental requirements[3]. With the advantages of non-pollution and renewable, it is recognized as an important development direction of the power industry in the future[4, 5].

When the distributed generation (DG) is connected to the distribution network, the single source system becomes multi-source, which will affect the power flow and direction of the original line, this will result in the network loss of the distribution network system to change[6]. The access capacity and position of the distributed power supply will have a certain influence on the network loss. However, an appropriate access capacity and position can reduce the system network loss. But the inappropriate access location and capacity will increase the network loss and reduce the stability of the system. Therefore, it is very important to optimally plan the access capacity and location of distributed power supply. In recent years, the optimization of distribution network is becoming a research hot spots, it has attracted the attention of academia at home and abroad.

Traditionally, particle swarm optimization (PSO) algorithm is used to solve the optimization DGs accessing to grid problem, in which the maximum benefit of power generation is taken as the objective function after the distributed generation is connected to the distribution network[7]. Although the optimal solution can be found for the distribution net-work with distributed power supply, the particle swarm optimization algorithm is slow to converge, and the population is prone to fall into the local optimum, so more iterations are needed to reach the optimal solution. In order to realize the optimal accessing of DGs to distributed electrical networks, In existing research works, some methods and improved optimal algorithms has been applied to distribution network expansion planning for optimizing the location and capacity of distributed generation. Based on the improved genetic algorithm, the multiobjective optimal allocation of distributed power supply is proposed in the literature[8]. Because of the deficiency of genetic algorithm in the speed of optimization, an improved genetic algorithm based on particle swarm optimization is proposed by combining particle swarm optimization with genetic algorithm. With the combination of the two methods, the algorithm has some advantages in improving the convergence speed and can find a more reasonable optimization solution with fewer iterations[9,10]. However, the combined program is more complex and the parameter setting is more complicated. A new evolutionary algorithm, genetic algorithm and simulated annealing algorithm are proposed to solve the problem of the optimal installation position of distributed power supply under the condition of minimum network loss[11]. Based on the bacterial colony optimization algorithm, the optimal allocation of distributed power supply is proposed,and the optimization model with minimal active power loss is established[12].However,the optimization mechanism of bacterial colony optimization algorithm is complex. It is difficult to find the optimal solution of high quality.

Motivated by the above considerations, in this work, a method based on improved PSO algorithm is proposed, which integrates the dynamic adjustment mechanism and chaos into the PSO algorithm, optimizes the configuration of distributed power supply, and adjusts the parameters reasonably. The variation and influence of voltage and line loss before and after distributed power supply access to distribution network are analyzed, and then an improved shrinkage factor particle swarm optimization (Constriction Factor Particle Swarm Optimal, PSO) algorithm combined with hybrid simulated annealing (Simulated Annealing, SA) algorithm is proposed. This paper optimizes the solution of the distributed power supply constant volume problem, and compares the calculation results with those of the basic particle swarm optimization algorithm and simulated annealing particle swarm optimization algorithm. The proposed algorithm has good convergence rate, good convergence and validity. The simulation results demonstrated that the improved particle swarm optimization algorithm has a certain practical significance in the fixed volume and location problem of DG.

The paper is organized as follows. Section 2 analyzes the effects of DG on Distributed network and power flow algorithms. Section 3 presents the improved particle swarm optimization algorithm and its optimization model. Section 4, the IEEE33 Node is used to take as a benchmark electrical net system, and the simulation results illustrate the effectiveness of the modified algorithm. The conclusion is given in section 5.

2. Influence of DGs on distribution network and power flow algorithm

The combination of distributed generation and distribution network is the key research direction in the future, including the advantages of reducing investment cost and loss, zero pollution, high efficiency, variety of energy, etc[13].

The influence of the access of distributed generation on distribution network mainly includes the following two aspects:

Before and after DG access, the distribution of power flow in the original distribution network will change, thus affecting the line loss of the whole distribution system. Reasonable access capacity can reduce the system line loss, if the access capacity is not appropriate, it may increase the system line loss.

DG access can improve the transmission margin of distribution system, increase the voltage level and support of the whole system, but when multiple DG access to the distribution network, especially affected by environmental factors, Its access will inevitably bring many uncertain factors to the stability of the system and reduce the reliability of the distribution network by [14].

For radiative distribution networks, the essence of the forward push back generation power flow calculation method is the Gao Si elimination method of node counter current number. The basic flow path is power forward and voltage generation: after known node voltage and the end power of the network, it is assumed that the voltage at all nodes is the same as the starting voltage. Then according to the known power of the end of the network, the power flow through each branch and the power loss of the line are calculated from the end of the network, and the total injection power of the starting node is obtained by iterative calculation to the starting end of the line. Then it is the process of voltage back generation: according to the injection power obtained, the power distribution and voltage distribution of each node are calculated from the beginning end to the tail end, and the convergent constraint condition is satisfied by continuously pushing back the power distribution and voltage distribution. The structure of distribution network is radiative, and the forward push back power flow algorithm is suitable for the structure. Because of the small influence of the impedance ratio and the fast convergence speed, the forward push-back algorithm is adopted in this paper.

Its basic principle includes two parts: forward push and back generation. When the network node load and the root node voltage are known, the branch power is calculated from the terminal node to the first node without calculating the voltage loss, and after the power of each branch is obtained, each branch is calculated from the head to the end of the network. Voltage loss and voltage at each node. So iterate and calculate repeatedly until the result meets the set accuracy.

The forward push back method takes the power supply point of the distribution network as the root node, which is divided into two parts:

Forward push and back generation. The specific algorithm steps are as follows:

Read the data of the distribution network, and determine the type of node access to the distributed generation;

Calculate the injection power of each load node; first assume that the node voltage of each node is invariable, and all initial reactive power is zero;

The line end voltage and node power are known, the power loss and branch power of the branch are calculated from formula (1) to (4), and the injection power of the root node is obtained by iteration.

According to formulas (5) and (6), the voltage distribution of each node is calculated from the voltage value of the line root node and the calculated power loss and branch power.

If not, repeat steps 2 to 4 until convergence conditions are met and output results.

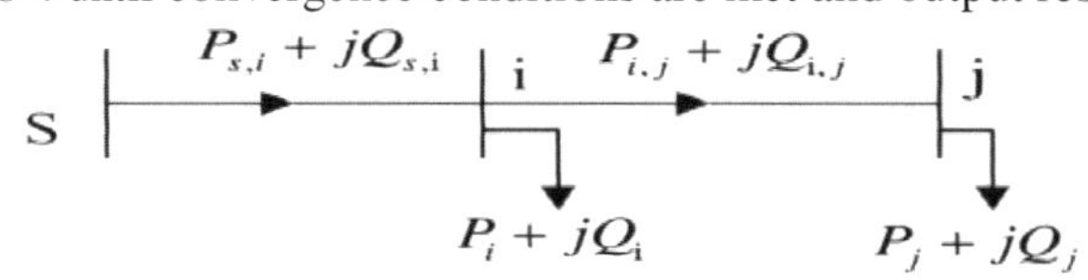

Figure 1. Radial distribution network.

The iterative formula of forward push-back power flow algorithm for distribution network is as follows:

The formula of power loss and power of the branch of node s and i.

$$\Delta P_{s,j}^{n} = R_{i,j}[(P_{i,j}^{n} + P_i)^2 + (Q_{i,j}^{n} + Q_i)^2] / (U_i^{n-1})^2 \tag{1}$$

$$\Delta Q_{s,j}^{n} = R_{i,j}[(P_{i,j}^{n} + P_i)^2 + (Q_{i,j}^{n} + Q_i)^2] / (U_i^{n-1})^2 \tag{2}$$

$$P_{s,j}^{n} = P_i + P_{i,j}^{n} + \Delta P_{s,j}^{n} \tag{3}$$

$$Q_{s,j}^{n} = Q_i + Q_{i,j}^{n} + \Delta Q_{s,j}^{n} \tag{4}$$

$P^n_{s,j}$ and $Q^n_{s,j}$ represent the resistance and reactance of branched i, j , $R_{i,j}$ and $X_{i,j}$ represent the Electrical Resistance and Impedance of the branched i, j , $P^n_{i,j}$ and $Q^n_{i,j}$ represent the active and reactive power rate of the branch circuit i, j , P_i and Q_i represent the node power.

The combination of DG and distribution network is the trend of the development of clean energy in the future. DG is a flexible power generation method, which can be installed directly near the load to supply the load, but also participate in the regulation of the distribution network power. Because the access of DG will bring all kinds of influence to distribution network, it is in practice. It is necessary to evaluate the performance and economy of distributed power supply access to the distribution network by appropriate evaluation methods or means in order to access the distribution network in the international distribution network.

The formula for calculating the voltage of node i is as follows：

$$\Delta U_i^n = \frac{P^n_{s,j} \times R_{i,j} + Q^n_{s,j} \times X_{i,j}}{U_s^n} \tag{5}$$

$$U_i^n = U_s^n - \Delta U_i^n \tag{6}$$

3. Improved particle swarm optimization algorithm and its optimization model

3.1 Particle swarm optimization algorithm

The basic idea of PSO is inspired by the population behavior of birds. It is based on the optimal search of velocity and position of each particle following the current optimal particle in the solution space. In the solution space of PSO,particles update their positions and velocities according to two optimal values. One is the optimal value of the particle itself so far, which is called the individual optimal value,and the other is the optimal value of the current population so far. PSO initialization produces random particles, each of which can have many dimensions. These particles then iterate through collaboration and competition to complete complex optimization searches in the solution space.

The updated formula is as follows:

$$V_i^{k+1} = w \times V_i^k + c_1 \times r_1 (p_{best} - X_i^k) + c_2 \times r_2 (G_{best} - X_i^k) \tag{7}$$

$$X_i^{k+1} = X_i^k + V_i^{k+1} \tag{8}$$

X_i^k is the rate of renewal of the k th of the i particle, the current position of the first particle k, w is the inertia factor,the reasonable inertia factor can well control the searching range of the particle, at the same time, It also plays a role in the global search ability of the population.Pbest represents the best search position to date (with the best fitness). Called particle optimal position (local optimal). Gbest represents all the best positions in the entire particle population so far(global optimality). r1 and r2 are random numbers, r1 ⊆[0,1] and r2 ⊆ [0,1]. They are used to adjust the orientation of updates. C1 and C2 are called learning factors.

3.2 Improved particle swarm optimization algorithm

The improved Particle Swarm Optimization (PSO) and optimization Model Particle Swarm Optimization (PSO) algorithm is a stochastic evolutionary algorithm based on intelligent search of bird model. It was first proposed by Dr.Kennedy and Dr. Eberhart. The flow and parameters of the algorithm are relatively simple. It has the characteristics of fast convergence for solving optimization problems [15]. Although the convergence rate of PSO is relatively fast, the lower convergence precision will make the population fall into local optimum. Therefore, in reference,an improved algorithm is proposed, which makes up for the deficiency of PSO in precision and local optimization. Simulated annealing algorithm (SA) is a stochastic combinatorial optimization method. When the initial temperature is high enough,it is asymptotically convergent with the decreasing temperature, and can converge to the global optimal value with probability 1. Because it accepts the difference solution with some probability, it has the ability to jump out of the local optimum.

In this paper, the shrinkage factor particle swarm optimization algorithm is combined with simulated annealing algorithm, and an improved particle swarm optimization algorithm is proposed. The steps of the algorithm are as follows:

Step 1: Random initialization of the position and velocity of the particle swarm to form a particle matrix;

Step 2:Calculate the fitness value of each particle, record the position and fitness value of each particle in the individual optimal solution of each particle Pi, record the position and fitness value of the optimal solution of all particles in Pg;

Step3: sets the initial temperature of simulated annealing to = f(Pg)/ln5;

Step4: calculates the fitness of each particle at the current temperature;

Step5: determines the global optimal substitution value by roulette method Pg, then updates the velocity and position of each particle;

Step6: calculates the fitness of each particle. The value of Pi and the value of Pg are updated;

Step7: is annealed $t_{k+1} = \Gamma t_k$, λ is the annealing constant;

Step 8: if the termination condition (maximum iteration number) is satisfied, stops the calculation, outputs the result,otherwise, returns the Step 4 to continue the calculation.

Among them, the fitting degree formula is as follows:

$$TF(P_i) = \frac{e^{(-(f(P_i)-f(P_g))/t)}}{\sum_{i=1}^{n} e^{(-(f(P_i)-f(P_g))/t)}} \tag{9}$$

Velocity and location of particle renewal:

$$V_{i,j}^{t+1} = \Psi[V_{i,j}^{t} + c_1 r_1 (P_{i,j} - X_{i,j}^{t}) + c_2 r_2 (P_g - X_{i,j}^{t})] \tag{10}$$

$$X_{i,j}^{t+1} = X_{i,j}^{t} + V_{i,j}^{t+1} \tag{11}$$

$$\Psi = 2/(2 - C - \sqrt{C^2 - 4C}) \tag{12}$$

Where is the contraction factor, it effects the search capability.

$$C = c_1 + c_2 \tag{13}$$

3.3 Optimal Planning Model of distributed Power supply

In this paper, the objective function with minimum loss is selected:

$$\min \ \mathrm{F}=\sum_{i=1}^{N-1} P_{loss,i} \tag{14}$$

Where $P_{loss,i}$ represents the line loss of the i branch, N represents the total number of nodes.

Among them, the current constraints are as follows:

$$P_i = P_{i-1} - R_{i-1}(P_{i-1}^2 + Q_{i-1}^2)/U_{i-1}^2 - P_{1,i} \tag{15}$$

$$Q_i = Q_{i-1} - X_{i-1}(Q_{i-1}^2 + X_{i-1}^2)/U_{i-1}^2 - Q_{1,i} \tag{16}$$

$$P_s = P_{load} - \sum_{i=1}^{N-1} P_{DG,i} + P_{loss} \tag{17}$$

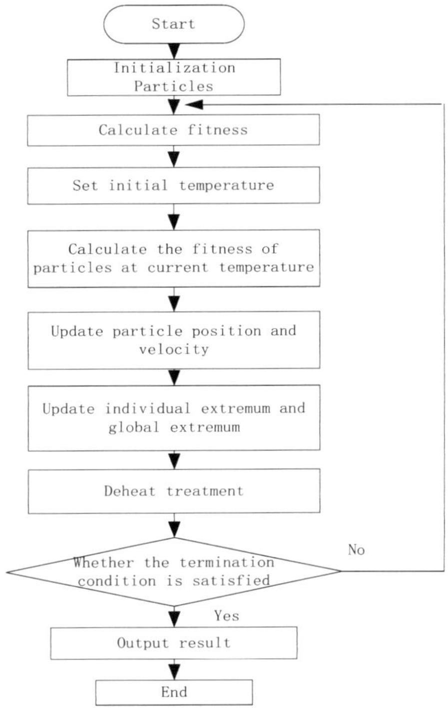

Figure 2. Flow chart of CFPSOGA algorithm.

In the formula: P_s is the initial power supply when the DG is not be connected, P_{load} is the usual total load power, and p loss is the total network loss of the system.

DG investment constraints include:

$$\sum_{i=1}^{N-1} P_{DG,i} \leq P_{DG,\max} \tag{18}$$

Constraints on the total installation capacity of distributed power access:

$$P_{DG,\min\leq} P_{DG,i} \leq P_{DG,\max} \tag{19}$$

P_{DG} is the DG capacity installed on the i node,; $P_{DG;\max}$ is the maximum installation capacity permitted by DG, and $P_{DG;\min}$ and $P_{DG;\max}$ are the minimum and maximum injection power of DG installation, respectively.System security operation constraints include: node voltage constraints:

$$\left|V_{i,\min}\right| \leq \left|V_i\right| \leq \left|V_{i,\max}\right| \tag{20}$$

$V_{i;\min}$ and $V_{i;\max}$ are represented as the upper and lower limits of the voltage allowed by the I node, respectively.Constraints of branch transmission power:

$$P_{s,j} \leq P_{s,j}^{\max} \tag{21}$$

4. Simulation and analysis

In this paper, the IEEE33 Node is used to take as a benchmark electrical net system, the structure of the node system is simplified as shown in figure 3.1:

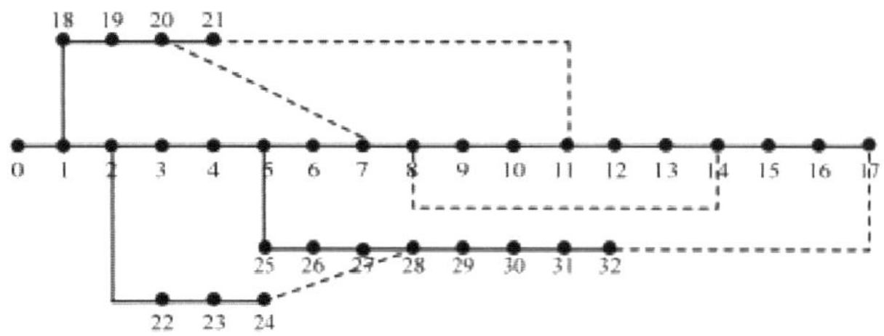

Figure 3. IEEE33 node Schematic diagram of distribution system structure.

When the improved PSO algorithm is used to solve the optimal network loss problem, the initial parameters are set as follows: the initial population size is 50; the maximum iteration number is 100; the learning factor is 2.05; the initial temperature is T = 10000; and the cooling rate is 0.5. In this paper, the distributed power generation is connected to the 10 (10)(16) (16) (24) (-1) node and the different algorithms are compared.

From the simulation results, the convergence speed of PSOSA is faster than that of PSO, in the early iteration, and at the latter stage, the optimal value found by PSOSA is better than that found by PSO, at the 67th iteration. However,the PSO falls into the local optimal value and needs more iteration to finally converge. It shows that PSOSA has better local and global search ability. At the same time, the active power line loss of distributed power distribution network obtained by PSOSA algorithm is smaller and has certain advantages, which shows that particle swarm optimization annealing algorithm is feasible in distributed power generation optimization planning. Moreover, it has better local search ability and global search ability than PSO algorithm,has stronger convergence and adaptability, and can search the optimal access capacity more quickly.

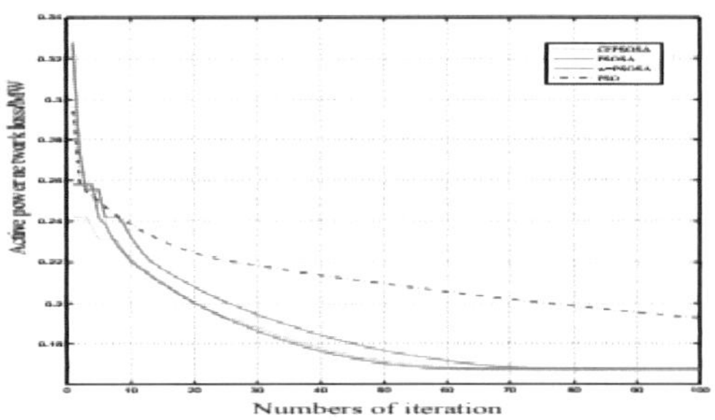

Figure 4. Comparison of three algorithms with PS.

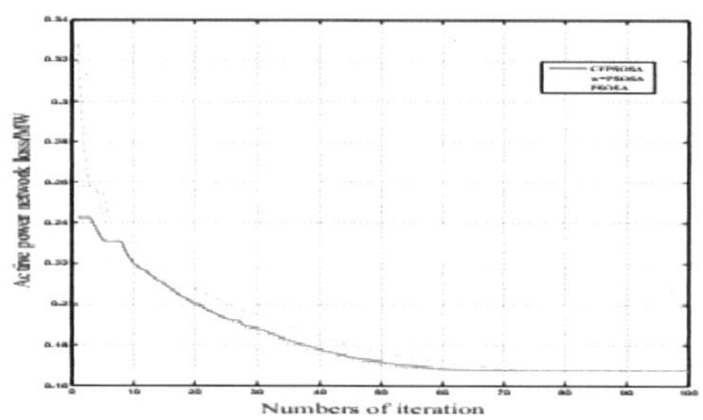

Figure5. Comparison Diagram of three algorithms.

At the same time, comparing the system active power network losses before and after DG access, Compared with the convergence of CFPSOSA and PSOSA, it can be seen that the contraction factor affects the search ability, and CFPSOSA shows better equilibrium local search ability and global search ability. The superiority of the CFPSOSA algorithm selected in this paper is verified. Table 1 shows the optimal access capacity of the node location calculated by the CFPSOSA algorithm.

Table 1. Access position and optimal capacity of distributed Power supply.

DG access position	DG optimal access capacity
10	0.06MW
16	0.06MW
24	0.42MW
31	0.15MW

The access position and optimal capacity of the distributed power generation are also connected to the distributed power supply at 10 16 24 31 nodes. For the PSOSA algorithm,the influence of inertial coefficient on the algorithm is compared. Figure 5 shows the influence of inertia coefficient on the searching range of particles and the global search ability. When w is large, the global search ability is strong, and when w is small, the local search ability is enhanced.

As can be seen from figure 6, when w = 0.729, in the iterative process, the solution will fall into a local optimum, it will have to jump out through the number of iterations, and,during the iteration, the rate of reduction of the active power line loss will also be relatively slow. That is, the ability to find the global optimal solution is relatively weak, and the search speed is slow. When the number of iterations is 80,the optimal solution is found. The global optimization can be found faster, when the number of iterations is 66, the optimal solution can be found, and the convergence speed is obviously faster than that of w= 0.729. Therefore, the value of inertial coefficient is directly related to the local and global search ability. The larger w, the faster convergence and the stronger ability to jump out of local search.

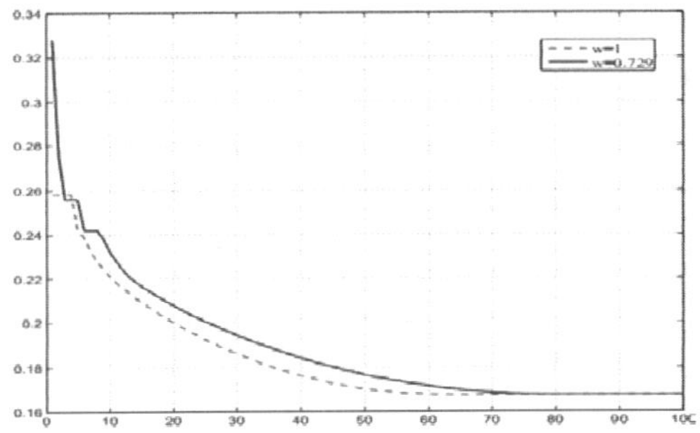

Figure 6. The influence of inertia coefficient w on search ability.

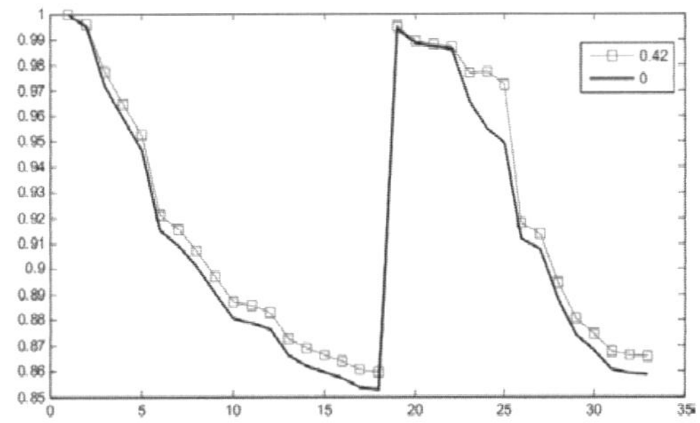

Figure 7. Voltage Distribution Diagram with access capacity of 0.42MW.

By using the forward and back generation power flow algorithm, it is concluded that the total active power load is3.715MW, the total reactive power load is 2.3Mvarand the active power network loss is 0.345 MW. When DG I connected, the line loss of the system calculated by the algorithm is 0.1674MW.Greatly reduced system line loss, reduced by 51.6 percent.

In fig.7, when a 24 node distributed power source with0.42MW capacity is accessed, it is compared with non-connected DG. It can be seen from the diagram that the power flow of the whole system will be affected when the distributed power supply is connected. The voltage distribution of each node is improved to some extent, especially in thevicinity of the access node.

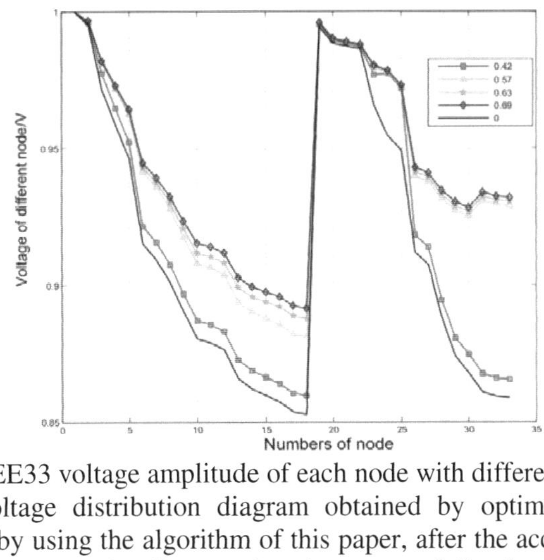

Figure 8. IEEE33 voltage amplitude of each node with different access capacity.

Fig.8 is the node voltage distribution diagram obtained by optimizing the configuration of distributed power supply by using the algorithm of this paper, after the access to DG node of 10H16 / 24 / 31 node. It can be seen from the diagram that the access of distributed power supply makes the voltage of distribution network system generally improve obviously.The amplitude of voltage rise is especially obvious, especially at the end node of the system, DG plays a very important role in the system voltage increase, so the DG access capacity increases. The voltage level near the DG access node is significantly increased, and when the DG access capacity increases, the overall voltage level of the system node will be improved.

5. Discussion and future trends

In this paper, based on the optimal network loss, a new method is proposed to solve the distributed electrical constant capacitive access point optimization problem by using the shrinkage factor particle swarm optimization algorithm based on the hybrid simulated annealing algorithm. The system loss is minimized after the distribution network is connected to DG. Compared with the basic particle swarm optimization algorithm and simulated annealing particle swarm optimization algorithm, the improved algorithm improves the traditional particle swarm optimization algorithm, such as slow convergence accuracy and easy to fall into local oscillation. The superiority of the proposed improved algorithm in convergence speed and finding the global optimal solution is verified, and the speed and accuracy of convergence to the global optimal solution are greatly improved. It is proved that the algorithm is effective for optimizing DG access to distribution network. At the same time, the example shows that DG can effectively improve the voltage distribution and reduce the line loss.

References

[1] Ji Yang, Ai Qian, Jie Da. The Development trend of distributed Generation Technology and Smart Grid Technology [J]. Grid Technology, 2010, 34 (12): 15-23.
[2] Liang Caihao, DuanXianzhong. Distributed generation and it-s influence on power system [J]. Power system automation, 2001, 25(12): 53-56.
[3] Qi peng Song, XiaoHuisong,Liang Yang. The influence of DG on distribution network planning [J]. Electrical Era, 2010474-75.
[4] H. Ouafa, K. Nadhir, S. Linda and B. Tarek, Optimal Power Flow with Emission Controlled using Firefly Algorithm, ICM-SAO Tunisia, 978-1-4673-5814-9/13, IEEE, (2013).
[5] A. Rezvani, M. Gandomkar, M. Izadbakhsh, and A. Ahmadi, Environmental/economic scheduling of a micro-grid with renewable energy resources, Journal of cleaner production, vol. 87, pp. 216-226, 2015.
[6] Wang Chengshan, Chen Kai, XieYinghua, et al. Siting and sizing of distributed generation in distribution network expansion planning[J]. Automation of Electric Power Systems, 2006, 30(3): 38-43.
[7] Liu Bo, Zhang Yan, Yang Na. Application of improved Particle Swarm Optimization algorithm in location and volume determination of distributed Power sources [J]. Journal of Electrical Technology, 2008, 23(2): 103-108.
[8] Wang Ruiqi, Li Ke, Zhang Chenghui, et al. Distributed Power Generation Planning based on Multiobjective chaotic Quantum genetic algorithm [J]. Power grid technology, 2011, 35(12): 183-189.
[9] Gandomkar M, Vakilian M, Ehsan M. A combination of genetic algorithm and simulated annealing for optimal DG allocation in distribution networks[C]. Proceeding of IEEE Canadian Conference on Electrical and Computer Engineering, 2005: 645-648.
[10] Shao Ke, Jiang Tiezheng. Optimal allocation of distributed Power supply based on bacterial colony Optimization algorithm [J]. Journal of Electric Power, 2014, 29(3): 201-205.
[11] Tang Nien, Xia Mingchao,Xiao Weidong, et al. Multiobjective expansion planning for distribution network considering multiple distributed generation and its stochastic characteristics [J]. Power system automation, 2015, 8: 008.
[12] Ji Yang, Ai Qian, Jie Da. The Development trend of distributed Generation Technology and Smart Grid Technology [J]. Power grid technology, 2010, 34(12): 15-23.
[13] Kennedy J, Eberhart R. Particle swarm optimization[C]. Proceedings of IEEE International Conference on Neural Net-works, 1995, 4: 1942-1948.
[14] Du Hinhui, Lu Xiaoxi, Xue Yingnan, et al. Location and capacity of distributed Power supply based on LD-SAPSO [J]. instrumentation 2015, 52(7): 118-122.
[15] Qu Xin, Wang Xiaojun, and Jing Han. Distributed Generation Planning Strategy for Distribution Network based on improved Particle Swarm Optimization algorithm [J]. Electrical applications, 2015, 2: 007.

IOP Conf. Series: Earth and Environmental Science **342** (2019) 012012 doi:10.1088/1755-1315/342/1/012012

Optimal Capacity Design of Independent Micro-Grid System for High Proportion of Wind-Solar Combined Cooling Heating and Power System with Energy Storage

Z L Yao[1], Z X Wang[1*], H Qin[2], L X Bao[3]
[1] Electronic Information and Electrical Engineering Department, Shanghai Jiao Tong University, 800 Dongchuan Road, Shanghai 200240, China
[2] Shanghai Huacheng Elevator Technology Co., Ltd, Shanghai 201111, China
[3] Shanghai Xilong Technology Co., Ltd, Shanghai 201517, China

*Corresponding author: wangzxin@sjtu.edu.cn

Abstract. This paper researches the optimal capacity design of independent micro-grid system for wind-solar combined cooling heating and power system with energy storage, puts forward a dispatching method based on expert time-judge for dynamic control of wind-solar direct permeability and a hierarchical design based on energy storage module. Because of the uncertainty of wind -solar system, this paper adopts fuzzy parameters to describe the system constraints and clarify the opportunity constraints, then optimizes the capacity design on fuzzy constraints and particle swarm optimization, analyses the operation results of optimal capacity allocation under dispatching method and hierarchical design of energy storage.

1. Introduction

Based on actual situation of China's energy supply and demand, the remote areas mainly develop diesel and gas power generation. The cost of power system is high and the reliability of system operation is low. Actually, most of the western regions are rich in renewable energy[1], which can be used to build renewable energy power generation system. Considering the cold and hot needs of users, an independent micro-grid system with multi-potential complementary renewable energy can be constructed.

This paper considers the independent micro-grid system for high proportion of wind-solar combined cooling heating and power system with energy storage[2-3] and this puts forward a dispatching method based on expert time-judge for dynamic control of wind-solar direct permeability to enhance the reliability of the system, a hierarchical design based on energy storage module to reduce the battery charge and discharge times, then, designs a dispatching strategy, get the optimal capacity of system configuration, and discusses the influence of high proportion of renewable energy penetration for a dispatching method.

2. System modelling

2.1. Joint Modeling of Wind and Photovoltaic Power Plants

From the single analysis of wind power and photovoltaic output, the wind power output has randomness, and the photovoltaic output has the phenomenon of work at day, stop at night. However, from the wind and solar power combined output, the coupling characteristics of wind and solar complementary output force reduce the fluctuation of output to a certain extent. and continuity of the output is guaranteed. Therefore, this paper regards wind and solar power combined output as a power source.

Due to the environment, climate, and other factors[4], the output of renewable energy is difficult to predict, so, the renewable energy output adopt fuzzy model. The actual value wind and solar power combined output is used as the fuzzy parameter to establish the fuzzy chance constraint[5], the risk is controlled by confidence level.

$$P_{w-pv} = \omega_i P^{forecast} \tag{1}$$

In the formula (1), $\omega_i (i = 1,2,3,4)$ is the proportional coefficient, and $\omega_i \in (0,1)$, $P^{forecast}$ is the predicted value of the wind-solar power combined output.

2.2. Energy storage system modeling

In order to reduce the times of charge and discharge of battery, the energy storage system is divided into the charged and the discharge groups according to the operating status in real time.

The remaining capacity of the battery at t is related to the remaining capacity of the battery at $t-1$, the battery output during $[t, t-1]$, and the self-discharge amount per hour[6].

When the battery works, the remaining capacity at the time t is:

$$S(t) = S(t-1)(1-\sigma) - \frac{P_s(t)}{\eta} \tag{2}$$

In the formula (2), $S(t)$ is the remaining capacity of the battery at the time t, $P_s(t)$ is the charging and discharging power of the battery at t, η is battery efficiency, σ is the ratio of the battery per hour.

Upper and lower limits of battery pack capacity:

$$S_{s\,min} \le S_s(t) \le S_{s\,max} \tag{3}$$

Upper and lower limits of battery pack power:

$$P_{s\,min} \le P_s(t) \le P_{s\,max} \tag{4}$$

2.3. Modeling of Cooling Heating and Power System

The cooling heating and power system uses a micro gas turbine as the core device. As the gas turbine generates electricity, the discharged high-temperature flue gas is used for cooling, heating and supplying domestic hot water through a lithium bromide absorption chiller[7-8].

In the operation of independent micro-grid, in order to improve the utilization rate of renewable energy, this paper adopts the operation mode of "power determine heating (cold)".

The upper and lower limits of the output of the cooling heating and power system:

$$P_{t\,min} \le P_t(t) \le P_{t\,max} \tag{5}$$

In the formula (5), $P_t(t)$ is the output power of micro gas turbine.

The upper and lower limits of the climbing rate of the cooling heating and power system:

$$R_t^{down}\Delta t \le P_t(t) - P_t(t-1) \le R_t^{up}\Delta t \tag{6}$$

In the formula (6), R_t^{down}, R_t^{up} are up and down climbing power speed.

2.4. Clear Equivalence Class Conversion of System Fuzzy Opportunity Constraints

The constraint can be converted by clear equivalence class method, the system power balancing is:

$$\mathrm{Cr}\{P_{w-pv} + P_s + P_t - P_l = 0\} \ge \alpha \tag{7}$$

In the formula (7), P_{w-pv} is the wind-solar power combined output, P_l is load power, α is confidence level.

According to the clear equivalence class analysis method, the clear equivalence class of system power balance under trapezoidal fuzzy parameters is:

$$(2-2\alpha)(\omega_3 P_{w-pv}^{fore}) + (2\alpha - 1)(\omega_4 P_{w-pv}^{fore}) + P_s + P_t - P_l \le 0 \tag{8}$$

In the formula (8), P_{w-pv}^{fore} is the wind-solar predicted power combined output.

3. System dispatching strategy

The control system regulates wind farm and other subsystems through the control line, and changes the dispatching strategy through the line switch. Combined Power Generation System is shown in Fig 1.

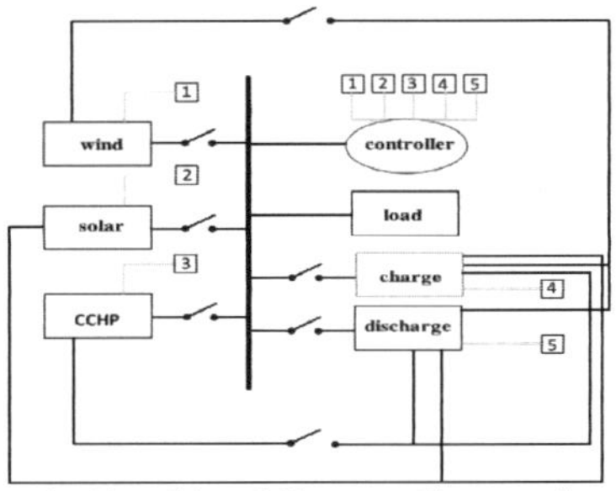

Figure 1. Combined Power Generation System

3.1. Expert time-judge

Different electric users have different load characteristics. Under the premise of ensuring the reliability of the system, the use of renewable energy during peak hours will be realized, and the phenomenon of abandoning wind and solar power should be reduced. First, peak hours and valleys hours should be divided by expert time-judge in different situations[9].

This paper adopts the time-judge plan proposed by tariff policy. The peak period of electricity consumption is 10:00-15:00 and 18:00-21:00, and the electricity valley is 00:00-7:00 and 23:00. - 24:00, weekdays are 7:00-10:00, 15:00-18:00 and 21:00-23:00.

3.2. Compound dispatching

The compound dispatching refers to a dispatching method in which two single-type dispatching modes are alternately operated. The control system uses S_s to determine which battery in the battery module is in a charging or discharging mode, thereby connecting the mode 1 and the mode 2, and the mode 1 refers to the wind-solar power priority input. The load is applied, and the direct penetration rate does not exceed the $\alpha\%$ of load; if there is more wind-solar power, the chargeable battery pack continues to be consumed; the remaining load is responsible for the combined heat and power supply system. Mode 2 means that the wind-solar power priority input, and the direct penetration rate does not exceed the $\alpha\%$ of load; the remaining load is supplied by the energy storage system and the cooling heating and power system. Determine the value of $\alpha\%$ according to peak, trough and time.

4. System capacity optimization

4.1. Economy

Total system construction cost measures economy C [10]:

$$C = C_w + C_{pv} + C_s + C_{CCHP} \tag{9}$$

In the formula (9), C_w, C_{pv}, C_{s1} and C_{CCHP} are the cost of wind power, photovoltaic, energy storage systems and cooling heating and power system, \$.

4.2. Power supply reliability

Load of power supply probability measures power supply reliability $LPSP$:

$$LPSP = \frac{\sum_{n=1}^{n_{max}} P_{lp}(n) \cdot 24/n_{max}}{\sum_{n=1}^{n_{max}} P_{LS}(n) \cdot 24/n_{max}} \tag{10}$$

In the formula (10), n_{max} means equal time periods numbers; $P_{lp}(n)$ is the average abandonment power of the first time of the day, kW; $P_{LS}(n)$ is the average load of the first time of the day, kW.

4.3. Renewable energy efficiency

Abandonment rate measures renewable energy efficiency A:

$$A = \frac{\sum_{n=1}^{n_{max}} P_a(n) \cdot 24/n_{max}}{\sum_{n=1}^{n_{max}} [P_{wf}(n) + P_{pv}(n)] \cdot 24/n_{max}} \tag{11}$$

In the formula (11), $P_a(n)$ is the average abandoned wind power in the first time of the day, kW.

4.4. System capacity optimization model

Convert the above three objective functions into joint satisfaction single objective function [11]:

$$\min\{\mu\} = \min\{\alpha_1\mu_1 + \alpha_2\mu_2 + \alpha_3\mu_3\} \tag{12}$$

In the formula (12), μ_1, μ_2 and μ_3 are satisfaction degrees after normalization of $C, LPSP$ and A, α_1, α_2 and α_3 the satisfaction weight chosen for preference based. The particle swarm optimization algorithm[12] is used to find the optimal solution of the system.

5. Case study

This paper selects a wind-solar combined cooling heating and power system with energy storage as the analysis object. Given the 24-hour predicted value of the capacity of a single wind turbine and the 24-hour predicted value of a single solar photovoltaic, the rated capacity of a single battery unit is 1.2 kWh and the rated power is 1.2 kW. The maximum capacity of the energy storage device is 30 kWh, the minimum capacity is 2 kWh, the initial capacity is 10 kWh, and the charging and discharging efficiency is 0.87. The discharge rate is 0.005. Given the upper and lower limit power and climbing rate of the cooling heating and power system, and the predicted value of the load for 24 hours, the confidence level of the system's fuzzy constraints is set at 75%. Fig 2 is 24h system load and actual output.

5.1. System capacity optimization results

According to the different wind-solar direct penetration plans, the optimal capacity allocation of the system is obtained by the method of system optimization allocation strategy in Section 3. The operating results of the system under different direct permeability are obtained in Fig. 3.

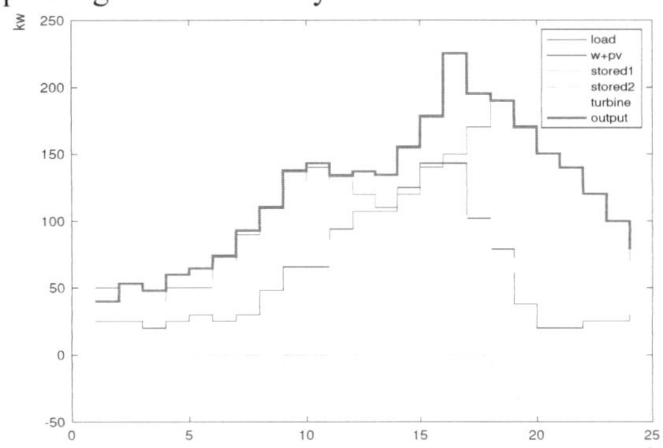

Figure 2. 24h system load and actual output

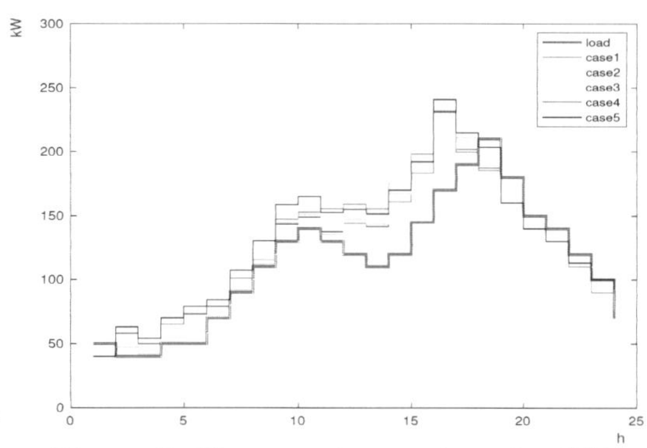

Figure 3. The operating results under different wind-solar direct permeability

From the point of view of the system operation results following the load, with the increase of the wind-solar direct penetration, the system operation deviation becomes larger and larger. There are two reasons: first, the higher the wind-solar direct penetration, the smaller the capacity demand of energy storage, so the stable system output function of energy storage system can't work; second, the increase of wind-solar direct penetration makes the residual load change more rapid, and the climbing power of CCHP system and energy storage system is limited, so the system is difficult to follow the load.

5.2. The influence of direct permeability of wind and solar

The influence of direct permeability of wind and solar is shown in Table1.

Table 1. The influence of direct permeability

Direct permeability	Economy	Reliability	Environment
$\begin{cases} low = 0.3 \\ norm - 0.45 \\ high = 0.75 \end{cases}$	1	0.17	0.5
$\begin{cases} low = 0.2 \\ norm - 0.35 \\ high = 0.65 \end{cases}$	0.6	0.36	0.4
$\begin{cases} low = 0.1 \\ norm - 0.25 \\ high = 0.55 \end{cases}$	0	0.39	0.3
$\begin{cases} low = 0.3 \\ norm - 0.45 \\ high = 0.65 \end{cases}$	0.9	0.16	0.467
$\begin{cases} low = 0.3 \\ norm - 0.35 \\ high = 0.65 \end{cases}$	0.3	0.26	0.433

Overall, with the increase of direct penetration of wind and solar power, the economy and environmental protection of the system are improved, while the reliability of power supply is reduced.

The economic improvement is due to the increase of direct wind-solar permeability, which makes fewer wind-solar input energy storage elements, the system's capacity demand for energy storage elements is reduced, and the load proportion of energy storage elements is reduced. At the same time, the increase of wind-solar permeability also reduces the output of the CCHP system, so the system cost is reduced and the system economy is improved. The improvement of environmental protection is mainly due to the improvement of utilization rate of wind photovoltaics.

The reasons for the decrease of power supply reliability are as follows: firstly, with the increase of wind-solar direct penetration, the charging and discharging time of energy storage system is prolonged, which leads to the increase of the probability of "load loss"; secondly, with the increase of direct penetration of wind and solar energy, the capacity demand of the system for the combined cooling, heating and power system decreases, so the system produces "power loss". The probability of load loss increase.

6. Conclusion

This paper researches the optimal capacity design of independent micro-grid system for wind-solar combined cooling heating and power system with energy storage, puts forward a dispatching method based on expert time-judge for dynamic control of wind-solar direct permeability and a hierarchical design based on energy storage module, combining with example simulation.

The operation results of the optimal system capacity under different direct permeability of wind and solar are obtained, and the influence of direct permeability of wind and solar on the system capacity planning is analysed. which has the guiding significance of the independent micro-grid system for high proportion of wind-solar combined cooling heating and power system with energy storage plan.

7. References

[1] Chauhan A, Saini RJR, Reviews SE. A review on integrated renewable energy system based power generation for stand-alone applications: configurations, storage options, sizing methodologies and control. 2014;38:99-120.

[2] Liang H, Zhuang WJE. Stochastic modeling and optimization in a microgrid: A survey. 2014;7(4):2027-50.

[3] Singh S, Singh M, Kaushik SCJEC, Management. Feasibility study of an islanded microgrid in rural area consisting of PV, wind, biomass and battery energy storage system. 2016;128:178-90.

[4] Sri-Jayantha SM. Smarter-grid: method to forecast electric energy production and utilization subject to uncertain environmental variables. Google Patents; 2013.
[5] Suganthi L, Iniyan S, Samuel AAJR, reviews se. Applications of fuzzy logic in renewable energy systems–a review. 2015;48:585-607.
[6] Bahmani-Firouzi B, Azizipanah-Abarghooee RJIJoEP, Systems E. Optimal sizing of battery energy storage for micro-grid operation management using a new improved bat algorithm. 2014;56:42-54.
[7] Friesth KL. Hybrid trigeneration system based microgrid combined cooling, heat and power providing heating, cooling, electrical generation and energy storage using an integrated automation system for monitor, analysis and control. Google Patents; 2017.
[8] Reddi KR, Li W, Wang B, Moon YJIJoSE. System dynamics modelling of hybrid renewable energy systems and combined heating and power generator. 2013;6(1):31-47.
[9] Mikkola J, Lund PDJE. Modeling flexibility and optimal use of existing power plants with large-scale variable renewable power schemes. 2016;112:364-75.
[10] Khan MJ, Yadav AK, Mathew LJR, Reviews SE. Techno economic feasibility analysis of different combinations of PV-Wind-Diesel-Battery hybrid system for telecommunication applications in different cities of Punjab, India. 2017;76:577-607.
[11] Erdinc O, Uzunoglu MJR, Reviews SE. Optimum design of hybrid renewable energy systems: Overview of different approaches. 2012;16(3):1412-25.
[12] Stoppato A, Cavazzini G, Ardizzon G, Rossetti AJE. A PSO (particle swarm optimization)-based model for the optimal management of a small PV (Photovoltaic)-pump hydro energy storage in a rural dry area. 2014;76:168-74.

Acknowledgments
This work was financially supported by Technical Standard Special Item of Shanghai Science and Technology Commission (18DZ2205700), Academician Expert Workstation Platform Project of Jinshan District (JSZJZ2018025G) and Industry-University-Research Cooperation Program of Minhang District, Shanghai (2017MH271).

IOP Conf. Series: Earth and Environmental Science **342** (2019) 012013 doi:10.1088/1755-1315/342/1/012013

Solar parabolic trough thermal energy output forecasting based on K-Nearest Neighbors approach

Jafferi Jamaludin[1] and Hiromitsu Ohmori[2]

[1] Higher Institution's Center of Excellence (HICoE), University of Malaya Power Energy Dedicated Advanced Center (UMPEDAC), Level 4, Wisma R & D, University of Malaya, Jalan Pantai Baharu 59990 Kuala Lumpur, Malaysia

[2] Department of System Design Engineering, Faculty of Science and Technology, Keio University, 3-14-1 Hiyoshi, Kohoku-ku, Yokohama 223-8522, Japan

E-mail: jafferi@um.edu.my

Abstract. Solar thermal output forecasting that is derived from the solar irradiance forecasting, is very much exposed to high forecasting error. This is because of the heavy dependence on the solar irradiance prediction accuracy that could be very low in certain situations owing to the high uncertainty in weather conditions. By considering this fact, this paper proposes to develop a solar thermal output forecasting model using measured solar irradiance data, instead of the predicted data. The proposed model applies K-Nearest Neighbors (K-NN) algorithm to generate 24-hour ahead forecasting data on solar thermal output from a solar parabolic trough system. For the purpose of illustrating the forecasting model performance, Kuala Lumpur, Malaysia is used as a case study, with PolyTrough 1800 model is selected as the solar parabolic trough collector under investigation. Simulation has been carried out using Matlab software to verify the effectiveness of the proposed K-NN-based forecasting model. The results show that the model is able to produce acceptable results in certain conditions.

1. Introduction

Concentrating solar power technology has rapidly grown in recent years owing to its ability to provide high temperatures in the range of 100°C to 450°C or even more. The technology basically focuses the sunlight at a receiver to achieve high temperatures. One of the most popular receivers is the solar parabolic trough concentrator. This concentrator offers several advantages which include high power density, high efficiency, modularity and versatility [1]. In addition, since it focuses on the direct radiation from the sun, a tracking system is required. The fact that this concentrator can reach high temperature range makes the solar parabolic trough system popular as a source of electricity since higher receiver temperature leads to higher efficiency and thus, results in better electricity generation.

Despite the strengths of the solar parabolic trough system, its solar thermal output is very much dependent on weather conditions. This introduces uncertainty in electricity generation which consequently causes power fluctuation, voltage stability risk and energy imbalance between supply and demand. As a means to reduce this uncertainty, solar irradiance forecasting has been widely employed [2]. The forecasting result is then used to estimate the solar thermal output. A number of studies have been conducted to forecast solar thermal output based on solar irradiance forecasting results. In [3], two Numerical Weather Prediction (NWP) models namely the Aerosol-based Forecasts of Solar Irradiance for Energy Applications (AFSOL) and the European Centre for Medium-range

Weather Forecasts (ECMWF) in combination with the Skartveit and Olseth global-to-diffuse model have been investigated to forecast the direct normal irradiance. The forecasting results are then converted to solar thermal output prediction based on the physical model of the parabolic trough solar thermal plant. In another study [4], by using the parabolic trough model as well, the solar thermal output forecasting is derived from the direct normal irradiance forecasting using the ECMWF together with the High-Resolution Limited Area Model (HIRLAM). [5] proposes the forecast technique of a one-day-ahead hourly thermal energy collection based on solar radiation forecasting by using three neural network models: feedforward neural network, radial basis function neural network and recurrent neural network. A model that combines the meteorological forecasts and thermal performances of a solar Fresnel power plant is presented in [6]. By feeding the model with the forecasted direct normal irradiance which is inferred from the global horizontal irradiance, the power output of the plant can be estimated.

Most of the studies found in the literature translate the solar thermal output forecasting from the solar irradiance forecasting. This approach is useful for short-term forecasting whereby information on the solar thermal output is needed a few minutes to a few hours ahead. However, forecasting of this type depends very much on the accuracy of the solar irradiance forecasting. On the other hand, for cases where the measured solar irradiance data is available from the previous days, forecasting of the solar thermal output can be based on the measured data, instead of the predicted data. Hence, the forecasting error of the solar thermal output can be reduced since the dependence on the solar irradiance forecasting accuracy has been removed. Taking this into account, this paper presents the 24-hour ahead forecasting model of the solar thermal output from a solar parabolic trough system based on the available solar irradiance data. The K-Nearest Neighbors (K-NN) technique is proposed to develop the forecasting model. Section 2 details the proposed K-NN-based forecasting model. Section 3 provides a description of the case study used to verify the performance of the proposed model. Section 4 presents the simulation results while Section 5 draws the conclusion.

2. Forecasting model

K-NN is attractive because of the simplicity and ease of implementation. Moreover, in certain circumstances, it outperforms the more complex algorithms in producing results with better accuracy. In this study, K-NN is used to forecast solar thermal output 24 hours ahead based on the assumption that unknown observations which have high similarity to known observations should belong to the same class or should produce almost the same outcomes.

For the proposed K-NN model, a database of solar thermal output with n observed data points is constructed based on the measured solar irradiance data obtained. Determination of the solar thermal output is done by first calculating the solar thermal collector efficiency using (1) below [7]:

$$\eta = \eta_0 - a_1 \frac{T_{in} - T_{am}}{G} - a_2 \left(\frac{T_{in} - T_{am}}{G}\right)^2 \quad (1)$$

Here, η is the collector efficiency, η_0 is the maximum collector efficiency, a_1 is linear heat loss coefficient in $Wm^{-2}K^{-1}$, a_2 is quadratic heat loss coefficient in $Wm^{-2}K^{-2}$, T_{in} is the inlet temperature in K, T_{am} is the ambient temperature in K and G is the solar beam irradiance in Wm^{-2}. By using the calculated collector efficiency, the solar collector output S_j is obtained using (2):

$$S_j = \eta_j G_j \text{ with } j = 1, 2, 3 \ldots\ldots, n \quad (2)$$

For the 24-hour ahead forecasting to take place, the previous 24-hour solar collector output data series is taken as the base data points. In order to assess the similarity between an observed data point in the database of the solar thermal output and a base data point, the distance metric between the two data points is calculated. In this study, Euclidean distance as given in (3) is used:

$$d(S_b, S_j) = \left[(S_b - S_j)^2 + (t_b - t_j)^2\right]^{\frac{1}{2}} \quad (3)$$

S_b is the base data point while t_b and t_j are the times when S_b and S_j occur respectively. By using (3), the distances between all base data points and all observed data points are determined. For each base data point, all calculated distances are sorted in ascending order so that the observed data points are ranked with the nearest points at the top of the order. These nearest points which are usually referred to as the nearest neighbors, are then assigned with different weights. The closer the nearest neighbors to the base data point, the higher the weight is. This results in closer nearest points to contribute more to the forecasting. Here, Gaussian weighting function is used as expressed below:

$$w(S_b, S_j) = exp\left(\frac{-d(S_b,S_j)^2}{\sigma^2}\right) \tag{4}$$

σ is the width of the Gaussian function which needs to be properly chosen. In order to calculate the output that corresponds to each base data point, (5) applies:

$$S_o = \frac{\sum_{j=1}^{k} w(S_b,S_j)S_j}{\sum_{j=1}^{k} w(S_b,S_j)} \tag{5}$$

S_o is taken as the forecasted solar thermal output which depends on the number of nearest neighbors considered as denoted by k in (5). It is a normal practice to select k that leads to the minimum forecasting error. As a means to estimate the forecasting error, it is commonly accepted to use percent mean absolute error (%MAE), mean absolute error (MAE) and root mean square error (RMSE) metrics as given below:

$$\%MAE = \frac{\sum_{i=1}^{N}|S_{m,i}-S_{o,i}|}{\sum_{i=1}^{N} S_{m,i}} \times 100\% \tag{6}$$

$$MAE = \frac{1}{N}\sum_{i=1}^{N}|S_{m,i} - S_{o,i}| \tag{7}$$

$$RMSE = \left[\frac{1}{N}\sum_{i=1}^{N}(S_{m,i} - S_{o,i})^2\right]^{\frac{1}{2}} \tag{8}$$

In (6)–(8), $S_{m,i}$ and $S_{o,i}$ are respectively the measured and forecasted solar thermal outputs at the ith hourly interval.

3. Case study

For the purpose of testing the forecasting model described in Section 2, a case study has been conducted. In this study, Kuala Lumpur which is the capital of Malaysia has been chosen. In order to execute the forecasting algorithm, solar thermal output from a solar parabolic trough system is estimated. In this study, the parabolic trough collector of model PolyTrough 1800 from NEP Solar AG [8] is used. Based on (1), the following equation is obtained.

$$\eta = 0.689 - 0.36\frac{T_{in}-T_{am}}{G} - 0.0011\left(\frac{T_{in}-T_{am}}{G}\right)^2 \tag{9}$$

where $\eta_0 = 0.689$, $a_1 = 0.36$ WK^{-1}m^{-2} and $a_2 = 0.0011$ WK^{-2}m^{-2} [8]. By using (9), the solar collector efficiency can be calculated. From the calculated efficiency, the solar thermal output can be achieved by using (2). Here, the solar beam irradiation can be obtained from any source that provides available measured solar irradiation data. In this study, the data is taken from the photovoltaic geographical information system (PVGIS) [9]. For the purpose of the illustrating the performance of the forecasting algorithm, the PVGIS data is assumed to be of measured values.

4. Simulation results

Simulation has been carried out using Matlab software to evaluate the effectiveness of the proposed forecasting model. The solar irradiance data taken from the PVGIS are between January 2016 and April 2016 for the case of Kuala Lumpur, Malaysia. For the purpose of calculating the solar collector efficiency using (9), it is assumed that T_{in} is 150°C and T_{am} is obtained from the weather data provided by PVGIS during the aforementioned period. The database of the observed data points comprises the solar thermal output data from 1 January 2016 until 7 April 2016 which are calculated using (2). As for the calculated solar thermal output data between 8 April 2016 and 14 April 2016, they are treated as the base data points to derive the 24-hour forecasting data on 9, 10, 11, 12, 13, 14 and 15 April 2016. Here, forecasting on 9 April uses the 24-hour calculated solar thermal output data on 8 April as the base data points. The same goes for the 10 April forecasting which utilizes the 9 April 24-hour calculated solar thermal output data as the base data points. This is the same for other days as well.

The simulation results are shown in figure 1 considering k equals 20 and σ equals 0.5. It can be seen in figure 1(a) that forecasting on 9 April and 10 April record significant errors which are 103% and 47% respectively in terms of the %MAE. This happens as a result of a considerable difference in terms of the amount and hourly pattern of solar thermal output between 8 April and 9 April as well as between 9 April and 10 April. Since the forecasting uses the previous 24-hour data from the day before as the base data points, the forecasting algorithm is not able to produce the desired prediction results. As a means to address this shortcoming, the previous 1-week data is used to generate the average 24-hour data to be utilized as the base data points. As an example, in order to carry out the forecasting for 9 April, the base data points take the average hourly data from 2 April to 8 April, instead of 8 April only. Figure 1(b) displays the simulation results. It can be observed that the %MAE for 9 April and 10 April have reduced to 81% and 25% respectively. Figure 2, figure 3 and figure 4 details the error comparison. Despite the error reduction for 9 April and 10 April, there is however an increase in the errors for 11 April and 12 April. Nonetheless, in terms of the overall performance, the previous 1-week average approach records lower overall average errors for the 7 days as presented in Table 1.

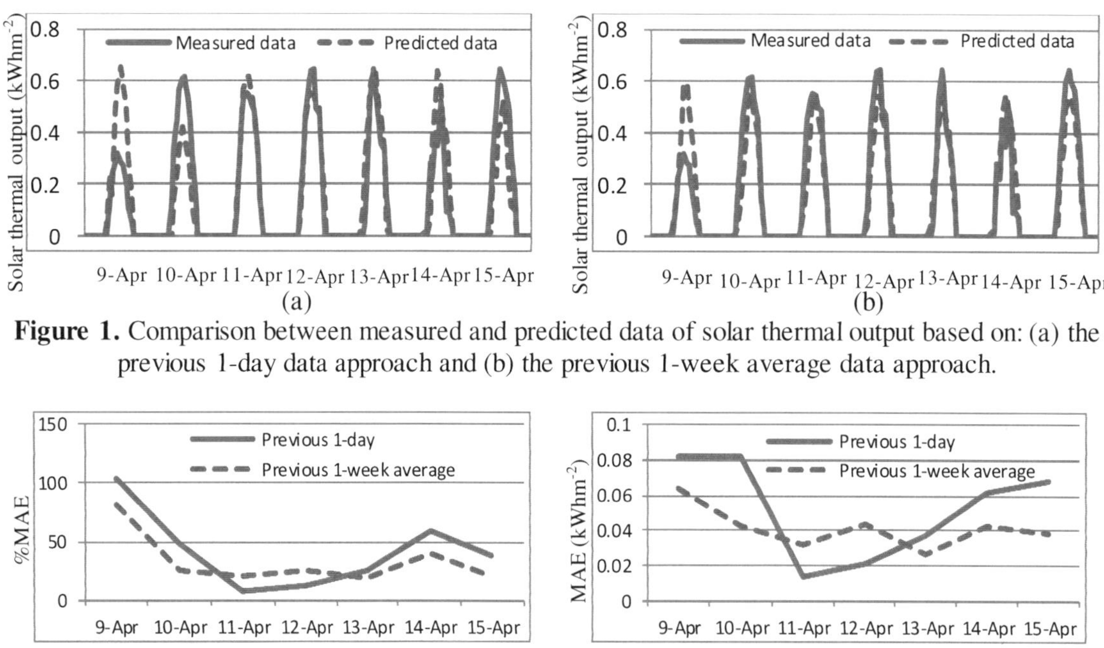

Figure 1. Comparison between measured and predicted data of solar thermal output based on: (a) the previous 1-day data approach and (b) the previous 1-week average data approach.

Figure 2. Comparion of %MAE.

Figure 3. Comparison of MAE.

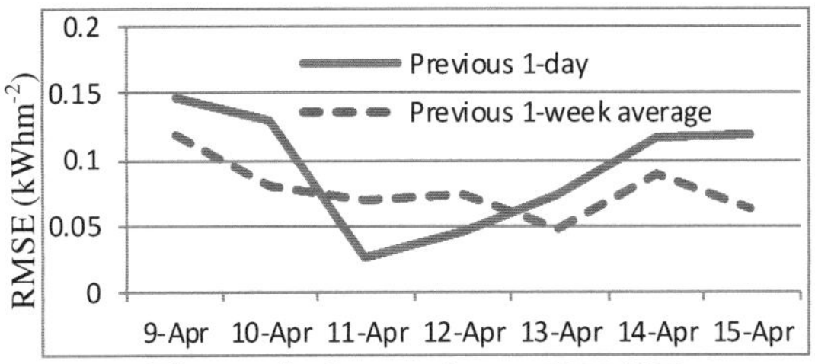

Figure 4. Comparison of RMSE.

Table 1. 7-day average errors comparison.

Error	Previous 1-day	Previous 1-week average
%MAE	41.9	32.9
MAE (kWhm^{-2})	0.052	0.041
RMSE (kWhm^{-2})	0.093	0.077

5. Conclusion

In this paper, K-NN approach has been proposed for 24-hour ahead solar thermal output forecasting from a solar parabolic trough system using measured data, instead of predicted data derived from solar irradiance forecasting. Two approaches to generate the base data points for the K-NN algorithm have been presented: previous 1-day and the previous 1-week average approaches. It has been demonstrated that the previous 1-week average approach produces forecasting results with lower errors in most instances. In addition, the previous 1-week average approach not only reduces the errors but also flattens the error curves. In conclusion, K-NN approach could be considered as a reliable approach to forecast solar thermal output without depending on the solar irradiance forecasting results. For future work, additional inputs influencing solar thermal outputs such as sun elevation, wind speed and rainfall will be considered to further improve the forecasting results.

6. References

[1] Hafez A Z, Attia A M, Eltwab H S, ElKousy A O, Afifi A A, AbdElhamid A G, AbdElqader A N, Fateen S-E K, El-Metwally K A, Soliman A and Ismail I M 2018 Design analysis of solar parabolic trough thermal collectors *Renew. Sustain. Energy Rev.* **82** 1215–60

[2] Law E W, Prasad A A, Kay M and Taylor R A 2014 Direct normal irradiance forecasting and its application to concentrated solar thermal output forecasting – a review *Sol. Energy* **108** 287–307

[3] Wittmann M, Breitkreuz H, Schroedter-Homscheidt M and Eck M 2008 Case studies on the use of solar irradiance forecasts for optimised operation strategies of solar thermal power plants *IEEE J. Sel. Topics Appl. Earth Obs. Remote Sensing* **1** 18–27

[4] Kraas B, Schroedter-Homscheidt M and Madlener R 2013 Economic merits of a state-of-the-art concentrating solar power forecasting system for participation in the Spanish electricity market *Sol. Energy* **93** 244–255

[5] Yona A and Senjyu T 2012 Thermal energy collection forecasting based on soft computing techniques for solar heat energy utilization system *Smart Grid Renew. Energy* **3** 214–21

[6] Rodat S, Tantolin C, Le Pivert X and Lespinats S 2016 Daily forecast of solar thermal energy production for heat storage management *J. Clean. Prod.* **139** 86–98

[7] Tzivanidis C, Bellos E, Korres D, Antonopoulos K A and Mitsopoulos G 2015 Thermal and optical efficiency investigation of a parabolic trough collector *Case Stud. Therm. Eng.* **6** 226–37

[8] Solar collector factsheet NEP Poly Trough 1800

[9] Huld T, Müller R and Gambardella A 2012 A new solar radiation database for estimating PV performance in Europe and Africa *Sol. Energy* **86** 1803-1815

Acknowledgments

This paper was produced as part of the research support program sponsored by The Hitachi Global Foundation and JST CREST Grant Number JPMJCR15K5 (Hayashi Team), Japan. The authors would like to express their appreciation to the sponsors for the support given.

IOP Conf. Series: Earth and Environmental Science **342** (2019) 012014 doi:10.1088/1755-1315/342/1/012014

Wind Energy Potential for Small-scale WEC Systems in Port Elizabeth

HC Tshimbiluni[1] and PY Tabakov[1]

[1]Durban University of Technology, Department of Mechanical Engineering, P.O Box 1334, Durban, 4000, South Africa

Email: happychristian42@gmail.com

Abstract. South Africa is the highest emitter of greenhouse gases in the continent, and 18th in the world. Electricity generation in South Africa is heavily dependent on fossil fuels for its energy needs. Hence, it is a necessity to shift to renewable energy sources for electricity production in order to mitigate carbon emissions. This paper discusses the potential for electricity generation through small-scale wind energy conversion (WEC) systems in Port Elizabeth that is considered to be South Africa's windiest city. A wind energy study was conducted through analyses of the local weather data. The study shows that Port Elizabeth has an annual wind power density of 138 W/m^2. The area studied has great wind power potential, especially for small-scale wind energy production. The Weibull method has been applied to predict wind power potential in Port Elizabeth, South Africa. Furthermore, the Weibull method was applied to model the probability function of the wind speed due to its user friendliness and flexibility.

1. Introduction

Renewable energy sources such as hydro, solar, wind, ocean thermal, geothermal and biomass energy have attracted significant focus worldwide due to their non-polluting and inexhaustible characteristics. Wind energy has proven to be the most suitable, inexpensive and effective source for power generation and is used extensively in various countries across Asia and European Countries that have the largest wind energy installed capacities also account for the largest wind turbine manufacturing market share [1]. These countries include Germany, China, India, Spain, Denmark, and the United States of America. The use of modern renewable energy sources such as solar and wind in Africa lags behind other regions. In 2011, the continent's estimated installed wind power capacity was 1.1GW. This constitutes less than 1% of the continent's total installed electricity generation capacity. The North Africa sub-region holds the majority of the continent's existing wind energy installed capacity. In sub-Saharan Africa, the biggest installed wind power capacity is in East Africa; wind power capacity of up to 186MW has been installed in seven countries within this sub-region [1]. Other sub-regions have made little or no progress at all such as Central Africa. The South African Wind Energy Association (SAWEA) states that 62% of South Africa's current energy consumption can be generated by wind power harvesting [2]. It is cheaper and environmentally friendly to generate electricity from wind compared to fossil fuels. Furthermore, the Renewable Energy Independent Power Producer Program (REIPPP) reported that the price of wind power generation has decreased by 42% and averages 74 cents per kWh in South Africa.

2. Background

The awareness of energy conservation is rapidly increasing as a result of global warming and increased electricity costs. The electricity prices in South Africa have risen in nominal and real terms throughout the past years. The regulator has agreed to this inflation increase for the next five years, causing concern among Eskom's customers [2]. Therefore, it is necessary to consider ways in which electricity can be accessible and affordable for the majority of households in South Africa. South Africa has an average wind speed of 5 m/s in some escarpment and coastal regions. However, this paper focuses on Port Elizabeth the windiest city in South Africa [1]. Table 1 shows the average wind speed of the study location.

Table 1.Average speed and measuring height in Port Elizabeth

Station	Average mean speed (m/s)	Height above ground level (m)	Height above sea level (m)
Port Elizabeth	5.5	10	60

The above information was obtained from the weather data recorded by the South African Weather Service (SAWS). Mean wind speeds equivalent of or above 4 m/s are highly suitable for small-scale wind power applications [4]. However, no small-scale WEC systems have been built for Port Elizabeth households and industrial areas in South Arica. The potential for wind power along the coastline has been somewhat neglected, hence it is necessary to produce more research work related to wind power.

3. Methodology

3.1. Site and Data Collection

Wind characteristics evaluations were conducted around South African windiest town, namely, Port Elizabeth. The compiled geographic data shows wind characteristics of a period of one year (January to December 2017). SAWS supplied the wind characteristics data, in addition, the data were measured at meteorological weather stations in Port Elizabeth. The measurements were taken at a standard height of 10 m above the ground.

3.2. Weibull Distribution Function

There are various distribution functions that can be applied in order to determine the probability density function of wind speed. However, in this study the Weibull distribution function (WDF) was used to establish the frequency curve of the wind speed probability function. The two-parameter Weibull probability density function is given as [4]:

$$f(v) = \frac{k}{c}\left(\frac{v}{c}\right)^{k-1} exp\left[-\left(\frac{v}{c}\right)^{k}\right] \quad (1)$$

$$(k>0, v>0, c>1)$$

Where v, is the observing speed, k is the Weibull shape parameter, c is the Weibull scale parameter and $f(v)$ is the probability of the observing speed. The Weibull parameters illustrate the wind potential of the site under evaluation. The shape parameter k indicates how 'peaked' the wind distribution is, whereas the scale parameter c indicates the 'windiness' of the site under consideration. The corresponding cumulative probability function of the Weibull distribution is given as [5]:

$$f(v) = 1 - exp\left[-\left(\frac{v}{c}\right)^{k}\right] \quad (2)$$

3.3. Parameter Estimation

Various researchers such as Dale & Ackerman [6] and Mulugetta & Drake [7] have used the maximum likelihood method to estimate the Weibull parameters. The shape parameter k can be estimated as follows [8]:

$$k = \left(\frac{\sigma}{\bar{v}}\right)^{-1.086} \quad (1 \leq k \leq 10) \tag{3}$$

Where σ is the wind speed variation, and $\bar{v}$ is the mean wind speed. The wind speed variance and mean wind speed can be determined based on the Weibull parameters as given below:

$$\bar{v} = \frac{1}{n}\sum_{i=1}^{n} v_i \tag{4}$$

And

$$\sigma^2 = \frac{1}{n-1}\sum_{i=1}^{n}(v_i - \bar{v})^2 \tag{5}$$

Where n & v_i are the number of observed non-zero wind speeds and wind speed, respectively. The scale parameter c is estimated by the following [8]:

$$c = (\bar{v})\left(\Gamma\left(1 + \frac{1}{k}\right)\right)^{-1} \tag{6}$$

Where $\Gamma(x)$ is the gamma function, calculated as:

$$\Gamma(x) = \sum_{0}^{\infty}(\mathrm{s}^{\mathrm{x}-1})(\mathrm{e}^{-\mathrm{s}})\mathrm{ds} \tag{7}$$

Integrating by parts the value of gamma function can be easily found as,

$$\Gamma\left(\frac{x}{2}\right) = \left(\frac{x}{2} - 1\right) \tag{8}$$

If x is even, otherwise, taking into account that,

$$\Gamma(0.5) = \sqrt{\pi} \tag{9}$$

And,

$$\Gamma(x+1) = x \tag{10}$$

The recursion formula should be used:

$$\Gamma(x) = (x-1)\Gamma(x-1) \tag{11}$$

And, used here by their kind courtesy.

3.4. Wind Power Density

The establishment of the wind power density per unit area is important when evaluating locations with wind energy potential. The wind power that flows at a speed v through a fin sweep area A increases as the cube of its velocity and is given by:

$$P_{(v)} = \frac{1}{2}\rho A V^3 \tag{12}$$

Wind power density (WPD) is normally expressed in Watts per square meter (W/m^2), this takes the frequency distribution of the wind speed into account. Furthermore, WPD depends on air density and the cube of the wind speed. Hence, it is generally considered a better indicator of the wind energy potential than wind speed.

$$P_w = \frac{1}{2}\rho c^3 \Gamma\left(1 + \frac{3}{k}\right) \tag{13}$$

Where ρ is the air density that can be determined as follows:

$$\rho = \frac{P}{RT} \tag{14}$$

Where P & T are the average pressure and temperature, R is the dry gas constant.

4. Wind Energy Analysis

4.1. Wind Characteristics

In this paper the recorded data for 2017 is used for analysis. The South African coastline's wind characteristics are evaluated across 12 months, i.e., on a yearly basis. Table 2 illustrates the mean monthly speeds from January to December in 2017.

Table 2. Annual average wind speed (m/s)

Location	Jan	Feb	Mar	Apr	May	Jun	Jul	Aug	Sep	Oct	Nov	Dec
Port Elizabeth	6.2	5.7	5.4	4.5	4.5	5.4	4.8	5.4	5.6	6.0	5.7	5.6

Table 2 illustrates the highest wind speed magnitudes in Port Elizabeth to be 6.2m/s. The wind speed range was at its lowest from April to July. The lowest wind speed magnitudes were observed to be 4.5m/s to 4.8m/s. Therefore, these recordings clearly indicate that the summer season is the windiest time of the year in the South African coastal region.

4.2. Weibull Probability Density Function

The Weibull probability density function (WPDF) is predominantly determined by the distribution function of the wind speed. The WPDF was applied in this study because it only consists of a few key parameters that can illuminate the characteristics of a wide range of wind speed data, such as hourly time-series wind speed data that may always be enormous. Hence, the WPDF is the most practical and simplest method to use in this application. Furthermore, it is easy to obtain the wind power potential and the economic viability, once the wind speed distribution is known. The shape parameter k indicates how 'peaked' the wind distribution is, whereas the scale parameter c indicates the 'windiness' of the site under consideration.

4.2.1. Weibull Probability Density Function in Port Elizabeth.

4.2.2. Port Elizabeth is classified as one of the windiest areas on the South African coastline [9]. The data were measured at a meteorological weather station in Port Elizabeth. The measurements were taken at a standard height of 10 m above the ground. Table 3 shows information related to the station location.

Table 3. Location of the PE meteorological weather station

Station	**Period (Months)**	**Latitude**	**Longitude**	**Height above sea level (m)**
Port Elizabeth	Jan-Dec	33.99	25.62	60

4.3. Wind Power Density

The wind power density in Port Elizabeth was determined by utilizing the wind parameters from the Weibull probability distribution. This statistical method is now considered to be a standard approach and is widely accepted for the evaluation of local wind probabilities. Equation (10) was applied to show wind power potential in the considered location, since it is generally considered a better indicator of the wind energy potential than wind speed.

4.3.1. Weibull Power Density in Port Elizabeth. Table 4 shows that the annual wind power density in Port Elizabeth (PE) varied from 77W/m^2 to 200W/m^2 throughout the year. The overall wind power density of the site is 138W/m^2. Hence, the power potential is 1209kWh/m^2 per year.

Table 4.Yearly wind power density in Port Elizabeth

Month	$V_{m\ (m/s)}$	$C_{(m/s)}$	**k**	$V_{maxE\ (m/s)}$	**WPD (W/m^2)**
January	6.2	6.900	3.081	8.117	**200**
February	5.7	6.380	2.910	8.259	**159**
March	5.4	6.044	2.782	7.343	**134**
April	4.5	5.052	2.715	6.191	**78**
May	4.5	5.019	2.033	7.114	**77**
June	5.4	6.093	2.351	7.917	**138**

July	4.8	5.413	2.382	6.991	**97**
August	5.4	6.088	2.439	7.782	**137**
September	5.6	6.311	2.674	7.778	**153**
October	6.0	6.764	2.547	8.492	**188**
November	5.7	6.298	3.264	7.291	**152**
December	5.6	6.243	2.876	7.501	**148**
Average	**5.4**	**6.051**	**2.672**	**7.565**	**138**

5. Conclusion

Port Elizabeth has an annual wind speed with maximum energy and wind power density of 7.565m/s and 138W/m^2, respectively. However, some months during spring and summer reach a peak of 200 W/m^2. The overall wind power density of the site is 138W/m^2. Therefore, Port Elizabeth has a wind power potential of 1209kWh/m^2 per year. Based on these findings Port Elizabeth has a great potential for wind power generation. Therefore, small-scale WEC systems would be suitable for areas around Port Elizabeth.

6. References

[1] Elena C, 2015. *The Wind Energy Industry Localization Roadmap in Support of Large-Scale Roll Out in South Africa. Urban-Econ.*

[2] Wiesegart K, 2011. *Options for the Establishment of a South African Wind Energy Centre (SAWEC).* Centre for Renewable &Sustainable Energy studies, Stellenbosch University, Private Bag X1, Matieland 7602.

[3] Elhadidy MA and Shaahid SM, 2000. *Parametric Study of Hybrid (Wind+Solar+Diesel) Power Generating Systems. Renewable Energy* **21** (2):129-139.

[4] Johnson, GL, 1985. *Wind Energy Systems*. Englewood Cliffs, NJ: Prentice-Hall.

[5] Akpinar ES, 2005. *An Assessment of Seasonal Analysis of Wind Energy Characteristics and Wind Turbine Characteristics.* Energy Conversion and Management **46**: 1848-1867.

[6] Dale JD and Ackerman MY, 1993. *Evaluation of the Performance of Attic Turbo-ventilators.* ASHRAE Transactions, **99**(1), 14-22.

[7] Mulugetta Y and Drake F, 1996. *Assessment of Solar and Wind Energy Resources in Ethiopia. II. Wind Energy*. Solar Energy **57**(4):323-334.

[8] Maatallah TES, Damouni AW and Nasrallah SB, 2013. *Wind Power Assessment and Evaluation of Electricity Generation in the Gulf of Tunis, Tunisia.* Sustainable Cities and Societies **6**: 1-10

[9] Masukume P, Makaka G and Tinarwo G, 2005. *An Assessment of Wind Energy Potential of the Amatole District in the Eastern Cape Province of South Africa*. University of Fort Hare, Department of Physics, Private. Bag X1314, Alice 5700, University of Venda, Department of Physics, P. Bag X5050, Thohoyandou, 0950.

Acknowledgement

First and foremost, I would like to acknowledge the guidance, wisdom and support I received from my supervisor, Prof Tabakov. I have been very fortunate to have the opportunity to work closely with a supervisor that has so much to offer and is willing to share it so freely. Furthermore, I would like to express my thanks to NRF for their financial support that enabled me to further my studies.

IOP Conf. Series: Earth and Environmental Science **342** (2019) 012015 doi:10.1088/1755-1315/342/1/012015

Data Management System for Photovoltaic Solar Environment

H Tak[1], D Lee[1], H Kim[1] and H G Cho[1]

[1]Department of Electrical Computer Engineering, Pusan National University, Busan, Korea

Email:hgcho@pusan.ac.kr

Abstract. As communication environments and Internet-of-Things technologies evolve, smart grid is being studied for collecting and predicting facility data in real time. In this paper, we propose a management system that collects the data of a photovoltaic system and predicts the amount of power generation. For this purpose, we designed an environment that can collect solar data, adjacent environment data, and facility information in real time. In addition, to expand the limits of previous power generation prediction studies, a power generation prediction experiment was conducted using recurrent neural network-based methods. The experimental results are compared with those of previous studies.

1. Introduction

Smart grid[1] is attracting attention as next-generation power grid for solving the problems of power demands, environmental pollution, and resource depletion. Studies on optimizing the smart grid network and predicting energy generation are actively being carried out. In this study, our aim is to predict solar power generation, which is the most widely used source of renewable energy [2].

Existing solar power generation prediction studies employ regression analysis or machine learning techniques. However, these studies have problems in that the length of the training data is fixed and the prediction performance is poor. Currently, deep learning techniques are widely used for prediction, but there are few cases of using neural network techniques to predict solar power generation. The recently developed neural network techniques can use historical data of various lengths as input data. As a result, they can flexibly cope with the length of training data while using complicated variables.

In this paper, we propose a management system that can measure the power data produced by a power facility and predict the solar power generation using information about the solar panels, related equipment, and environment. To do this, we describe the database design, data accumulation method, and solar power generation forecasting model that are necessary for constructing the management system. In addition, an optimized prediction model using three recurrent neural network (RNN) methods and environmental data sets was evaluated experimentally.

2. Related Work

Previous studies on smart grid and solar power generation have been conducted. Hosoda and Namerikawa [3] clustered solar power data using the type of weather, and predicted power generation using a regression equation for the grouped data. Yona et al. [4] modeled solar radiation, temperature, atmospheric solar radiation, and humidity over 24h period using artificial neural networks, RNNs, and predicted solar power generation accordingly. Gensler, Henze, and Sick [5] predicted solar power generation using a long short-term memory (LSTM) network, which is a type of RNN.

In the system described in this paper, data from the solar module and environmental information are collected using sensors. Hence, we also reference a study using environmental sensors [6].

3. Management System Architecture

The proposed management system consists of a data extraction module, data analysis module, and solar power generation prediction module. The data extraction module measures the data of the solar module, environmental sensor, and DAB converter. The measured data are sent to a database over Wi-Fi. The data analysis module calculates data statistics and determines abnormalities using past data. The solar power generation prediction module predicts power generation using past data and RNNs.
Figure 1 shows the architecture of the proposed system. As the data are accumulated, the prediction module predicts the amount of power using a model and data of various length. The prediction results are stored in the database. In the proposed system, it is possible to check the data measured at each facility, perform statistical analysis, and evaluate the results of power generation prediction.

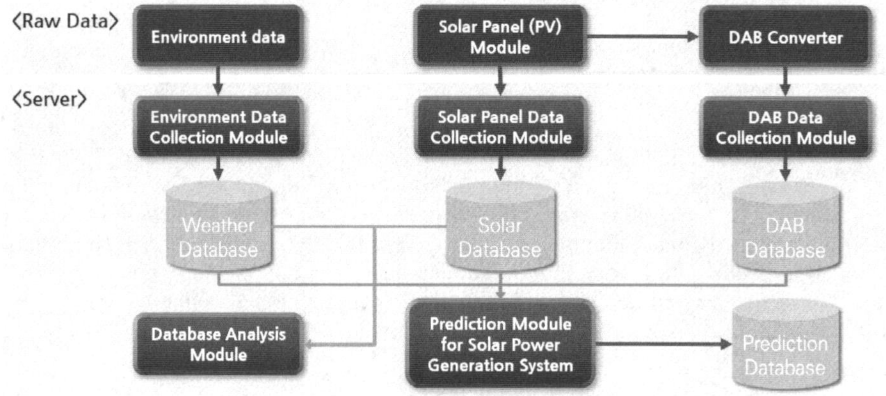

Figure 1. Architecture of the proposed data management system

4. Database Construction and Data Gathering

In the data management system proposed in this paper, solar module, DAB converter, and environmental sensor data are collected. The procedure is shown in Figure 2. Each measurement module consists of an Arduino and Wi-Fi shield. The measured data are stored in the database remotely using PHP. The data generated from the solar panel are stored in the Solar Database, the data measured in the environmental sensor is stored in the Weather Database, and the data generated in the DAB converter is stored in the DAB Database. Each data is stored in the server's database after time and place information is added.

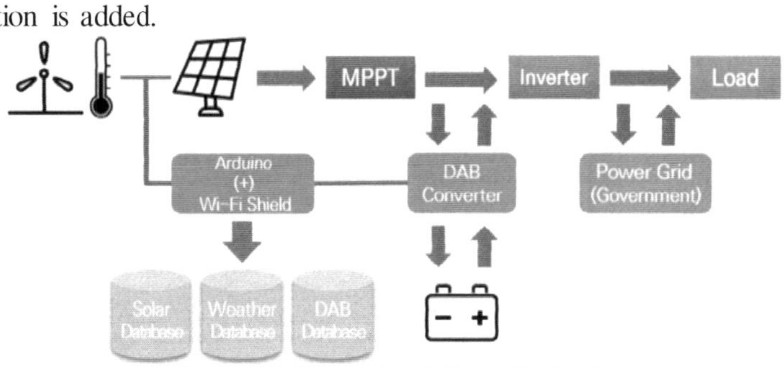

Figure 2. Example of Data Gathering

5. Prediction System for Power Generation System

For solar power generation prediction, we used the mean absolute percentage error (MAPE) and mean squared error (MSE) to evaluate the forecast results. MAPE and MSE are obtained by calculating the difference between the predicted result and the actual result using the absolute value or square and then calculating the average value as follows:

$$\text{MAPE} = \frac{100\%}{n} \cdot \sum_{i-1}^{n} \left| \frac{Y_i - \widehat{Y_i}}{Y_i} \right|$$

$$\text{MSE} = \frac{1}{n} \sum_{i=1}^{n} (Y_i - \widehat{Y_i})$$

Here, n is the number of data used in the evaluation, Y_i is the actual value, and $\widehat{Y_i}$ is result of prediction. To optimize the solar power generation prediction module, four weather factors that affect power generation were selected: temperature, humidity, solar irradiance, and duration of sunshine. For the prediction experiment, we designed five data sets shown in Table 1 using different combinations of factors. A comparison experiment was conducted using three RNN models: a basic RNN, a LSTM network and a gated recurrent unit (GRU), which is a network with excellent learning ability for time series data.

Table 1. Data Sets used for the Power Generation Prediction Model

Data Set	Input Data Parameter
1	Photovoltaic power generation
2	Photovoltaic power generation, solar irradiance, and temperatures
3	Humidity and duration of sunshine
4	Solar irradiance and temperatures
5	Solar irradiance, temperatures, humidity, and duration of sunshine

6. Experiment

In this study, we trained the three RNN models using historical data that varied in length from one hour to three days. If the data exceeded three days in length, there was no fluctuation in prediction rate, so longer periods were not evaluated. Figures 3–7 show the variation in MSE and solar power generation prediction results for each RNN and various lengths of historical training data for the data sets listed in Table 1.

There are differences in the results for each data set, but the error rate generally decreases when data for the past 24 h are used. However, the model overfit the data if historical data of a certain length are used. The experimental results show that the LSTM networks trained on 24 h of historical data obtain the best results. Figure 8 shows that the error rate is lowest when predictions are made by inputting temperature, humidity, solar irradiance, and duration of sunshine.

In existing studies, the prediction results are evaluated using MAPE or MSE. However, because the previous models and proposed model cannot be tested on the same data, the results reported by the authors on their own data are shown. The results are compared in Table 2, which shows that the error rate of the proposed prediction model is low.

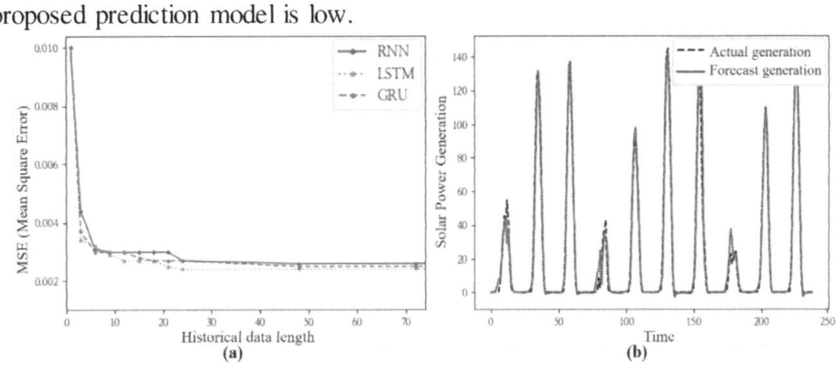

Figure 3. Prediction results for data set 1. (a) MSE with respect to the length of the historical training data. The error rate for LSTM with 24h of training data is the lowest (0.0024). (b) Forecasted results of LSTM (trained on 24h).

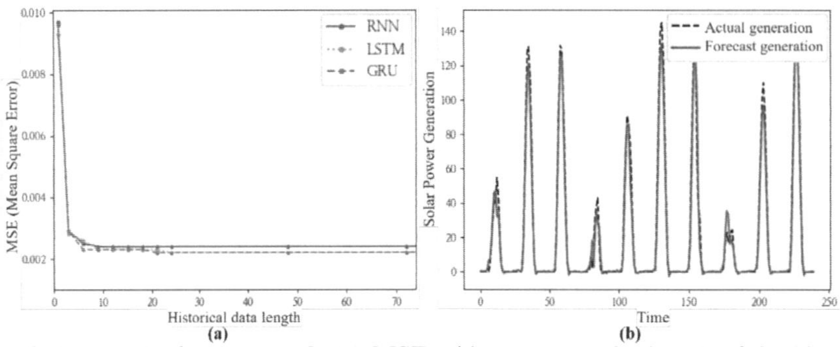

Figure 4. Prediction results for data set 2. (a) MSE with respect to the length of the historical training data. The error rate at LSTM with 24h of data and GRU with 21h of data are the lowest at 0.0022. (b) Forecasted results of LSTM.

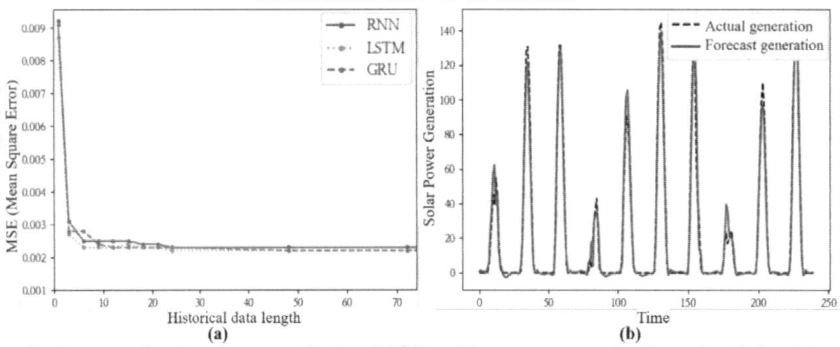

Figure 5. Prediction results for data set 3. (a) MSE with respect to the length of the historical training data. The lowest error rate (0.0022) was obtained by LSTM with 24h of data and GRU network with more than 24h of data. (b) Forecasted results of LSTM.

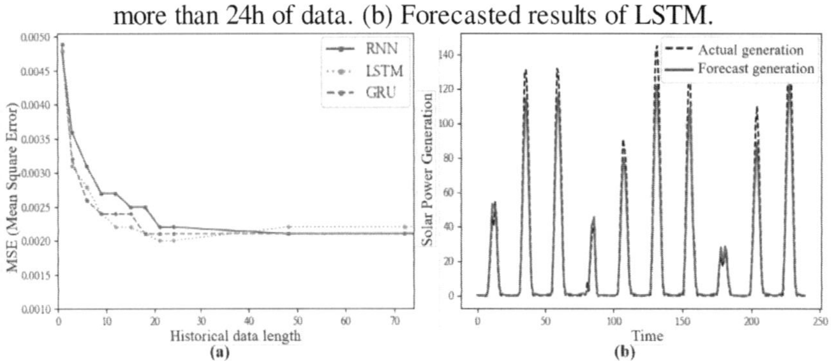

Figure 6. Prediction results for data set 4. (a) MSE with respect to the length of the historical training data. The error rate for LSTM with 24h of data is the lowest (0.002). (b) Forecasted results of LSTM.

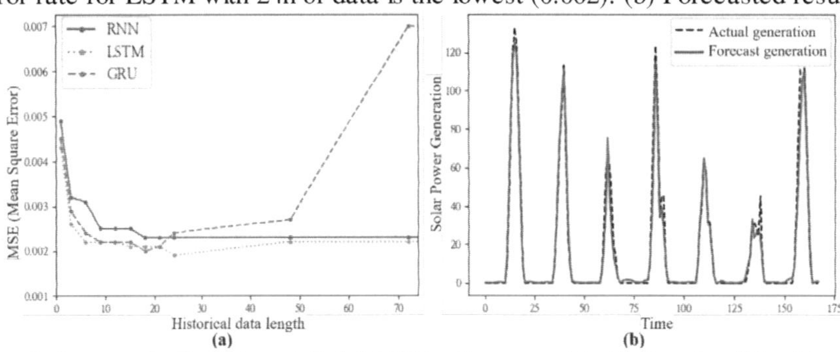

Figure 7. Prediction results for data set 5. (a) MSE with respect to the length of the historical training data. The error rate of LSTM with 24h of data is the lowest (0.0019). (b) Forecasted results of LSTM.

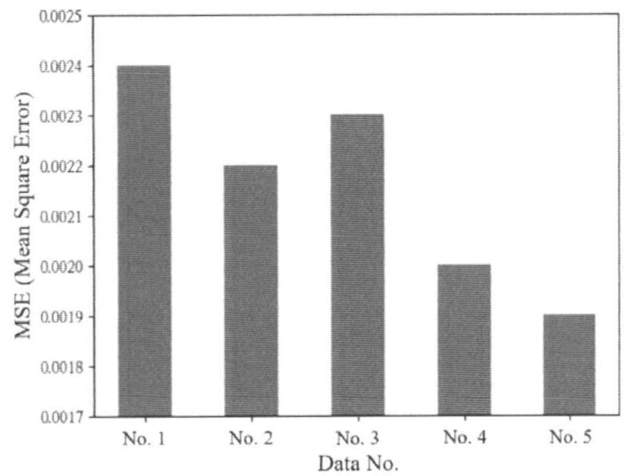

Figure 8. Comparison of the prediction error rate of LSTM with 24 h of past data on each data set. In data sets 2 and 3, the duration of sunshine is considered, but overfitting occurs and the error rate increases. Using the temperature and solar irradiance data together lowers the error rate with respect to the use of past photovoltaic power generation data alone. The lowest error rate is obtained when all four weather factors are used.

Table 2. Performance Comparison with existing Photovoltaic Power Research

Author	MAPE	MSE
Hosoda and Namerikawa [3]	13.87	N/A
Yona et al. [4]	14.97	N/A
Gensler, Henze, and Sick [5]	N/A	0.0049
Proposed Model	7.29	0.0019

7. Conclusion and Future work

In this paper, we proposed a data management system for smart grids. To do this, we introduced the modules needed to implement such a system. In addition, a prediction model based on an RNN was proposed and evaluated by experiments on historical solar power generation data of different lengths. LSTM was found to have excellent prediction performance, and it was confirmed that all four evaluated weather variables (temperature, humidity, solar irradiance, and duration of sunshine) were influential. In the future, we will implement an algorithm for optimal power allocation using the prediction model proposed in this study.

8. Reference

[1] Sung D, Song N, Go G, Cha J, Bae G and Jang H, 2013, *Communications of the Korean Institute of Information Scientists and Engineers*, **31**, 10
[2] Lee M and Kim S, 2013, *Journal of Korean Institute of Information Technology*, **11**, 181
[3] Hosoda Y and Namerikawa T, 2012, *SICE Annual Conference*, 119
[4] Yona A, Senjyu T, Saber A T, Funabashi H, Sekine H and Kim C, 2007, *Intelligent Systems Applications to Power Systems*, 1.
[5] Gensler A, Henze J and Sick B, 2016, *IEEE international conference on systems, man, and cybernetics (SMC)*, 2858.
[6] Ausilio A, 2012, *Behavior research methods*, **44**, 305.

Acknowledgments

This research was supported by Basic Research Laboratory through the National Research Foundations of Korea funded by the Ministry of Science, ICT and Future Planning (NRF-2015R1A4A1041584).

IOP Conf. Series: Earth and Environmental Science **342** (2019) 012016 doi:10.1088/1755-1315/342/1/012016

Application of Hybrid Model Predictive Control (Hybrid-MPC) to Microgrid with High Penetration of Renewable Energy

Zhichao Liu, Changhong Deng* and Yahong Chen

School of Electrical Engineering and Automation, Wuhan University

dengch@whu.edu.cn

Abstract. Owing to the cumulative impacts of stochastic forecast errors (SFE) of renewable energy generation, energy storage system (ESS) may be forced to quit operation owing to SOC (state of charge) deviation. Aimed at solving this problem, an online hybrid model predictive control (Hybrid-MPC) based strategy is proposed. Hybrid-MPC consists of two hierarchies: one is decreasing horizon rolling optimization which is specially for handling small SFE and the other is heuristic cooperative control which is designed to tackle large SFE. Besides, feedback correction is applied to deal with SOC deviation and maintain the normal operation of ESS. The strategy is tested in a self-developed program MG-ROS. Results demonstrate the effectiveness of the proposed strategy.

1. Introduction

MICRO GRID (MG) is generally composed of variable speed pumped storage unit (VSPS), energy storage system (ESS), photovoltaic (PV) cells and wind turbine (WT) generators. Day-ahead energy management system (EMS) combined with real-time droop control is a common method for the stable and economic operation of MG. Results of day-ahead optimal optimization are utilized as the set points for droop curves of VSPS and ESS [1]. And the purposes of adopting droop control are to instantaneous balance power supply and demands and compensate for the stochastic forecast errors (SFE) of renewable energy sources (RESs) generation and load demands [2]-[3].

In fact, applying droop control, SFE of RESs generation and load demands will be shared by VSPS and ESS. Consequently, the deviation between actual power and day-ahead schedule power for them [1] will be generated. Then energy error, i.e., state of charge (SOC) deviation of ESS, which is induced by the product of power error and time, is therefore produced by iterative calculation and continues increasing. SOC value is a critical factor to determine the operation status of ESS. The larger the deviation of SOC is, the larger the probability of over charge and over discharge of ESS is. If this happens, ESS will be forced to limit its power or even off-the-line by battery management system. Then operation stability of ESS and MG is reduced. Therefore, handling SOC deviation is extremely important for the normal operation of MG [4].

The objective of this paper is to propose an EMS strategy to cope with SOC deviation and maintain the normal operation of ESS. Recently, numerous models have been applied to MG operation optimization to alleviate the impacts of SFE of RESs generation, for instance, stochastic economic dispatch and robust optimization [5]-[6], etc. However, if penetration of RESs is high, the performance of these algorithms is not satisfactory enough. More pertinent strategy is thus needed.

Model predictive control (MPC) based online EMS approach has good robustness in dealing with uncertainty, therefore is a very promising an effective solution to above problems [7]-[8]. However, the main disadvantages of MPC are complex, computationally demanding, and that sometimes it has to apply decomposition techniques [7]. To solve this, a Hybrid-MPC method is proposed. The significant differences between Hybrid-MPC and MPC are: (i) tedious repeated ultra-short prediction is removed; (ii) time horizon for closed-loop rolling optimization is dynamically changing; (iii) high computation complexity for state equations of MG is avoided; (iv) day-ahead schedule is utilized as the basic reference command to guarantee global optimal; (v) real-time closed-loop feedback correction is applied to adjust SOC; (vi) under small SFE, time-varying rolling optimization is applied; (vii) under large SFE, heuristic control is implemented [9].

The main contributions are as follows.

1) An online closed-loop Hybrid-MPC based EMS strategy is proposed to cover the negative effect of SFE on SOC of ESS.

2) Decreasing horizon rolling optimization for handling small SFE is employed, while heuristic control to cope with large SFE is utilized.

3) Closed-loop feedback correction is applied to change the operation status of MG in time when necessary thus keeping SOC of ESS always within desirable limits.

The remaining sections are outlined as follows. Section 2 proposes the Hybrid-MPC based EMS strategy. In section 3 numerical case studies are presented. In section 4, conclusions are drawn.

2. Hybrid-MPC Based EMS Strategy

To cope with the problem of SOC deviation presented in literature [4] and maintain the normal operation of ESS, Hybrid-MPC EMS strategy is proposed. Hybrid-MPC consists of two hierarchies, time-varying rolling optimization (RO-MPC) is utilized to cope with small SFE and online heuristic control (HC-MPC) is applied to handle large SFE, respectively.

2.1. RO-MPC Based EMS Strategy

An appropriate errors threshold ε is set in the first place, if SFE of net load $\leq \varepsilon$, they are considered as small errors, otherwise, regarded as big errors.

For traditional MPC, algorithm is applied based on the current state of MG and ultra-short predictions of net load. Rolling optimization [10]-[11] is conducted over the fixed time horizon. This open-loop optimization problem is solved online. The first element of the optimization results (control sequence) is implemented to act on the controlled object. Then time horizon moves forward to next interval. The process is repeated again based on the latest measured information in the next sampling period.

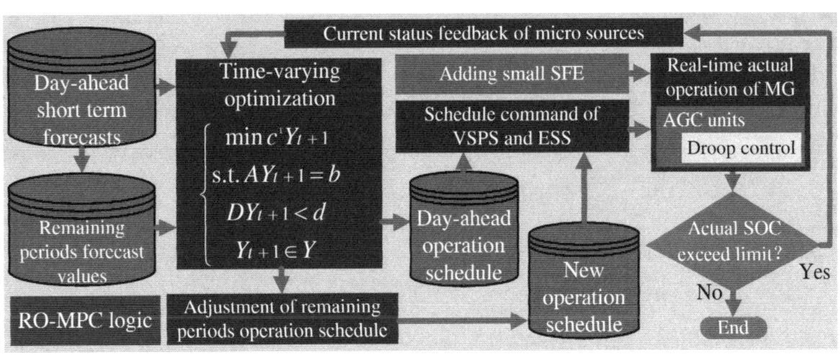

Figure 1. RO-MPC based EMS strategy.

For RO-MPC, time horizon for rolling optimization is decreasing with time. Calculation for time-varying rolling optimization is based on the day-ahead forecast value of net load. Real-time feedback of SOC is utilized for closed-loop correction. Implementation process of RO-MPC is briefly introduced as below. If SFE is small and actual SOC is within the normal range then operating commands of VSPS and ESS are set according to their day-ahead schedule. Thus operation economy can be ensured. However, if actual SOC exceeds the range then RO-MPC is applied. Current value of SOC is taken as initial calculation condition and day-ahead forecast values of net load for remaining periods are taken as input, CPLEX solver is called to change the operation schedule of remaining periods, as shown in figure 1. Thanks to the mechanism of SOC feedback, new schedule will adjust SOC towards opposite direction and change it gradually to normal range.

Equation of time-varying optimization is:

$$\min c^{\mathrm{T}} X_{t+1} \text{ s.t. } AX_{t+1} = b, DX_{t+1} < d, X_{t+1} \in X \tag{1}$$

Optimization results:

$$X_{t+1}^{\mathrm{T}} = \{p_i^{+}[t+1], \ldots, p_i^{+}[\mathrm{T}], p_{ess}^{+}[t+1], \ldots, p_{ess}^{+}[\mathrm{T}]\} \tag{2}$$

Feedback correction:

$$\Delta X_{t+1} = f_X[\mathrm{SOC}(t)] \tag{3}$$

Typically, the optimization model is formulated as a large scale mixed integer nonlinear programming (MINLP) problem. Objective function of MG, a set of equality and inequality operation constraints, and integer and binary optimization variables of micro sources are generally included. The objective function generally includes: 1) startup and shutdown cost of VSPS; 2) nonlinear quadratic consumption cost of VSPS; 3) charge-discharge and degradation penalty cost of ESS. Operation of VSPS is subject to a series of constraints, for instance, output power range constraint, maximum available power constraint, ramp rate constraint and minimum on time and off time constraint. To meet the instantaneous power balance requirement of MG while power of net load fluctuates, minimum standby rotation reserve constraint must be satisfied as well. Operation of ESS is also subject to a series of constraints such as charge and discharge power constraint and upper and lower SOC range constraint. Detailed constraints are not the focuses of this study, they are not listed and can be seen in references [1] and [5]-[6].

2.2. Online HC-MPC Based EMS Strategy

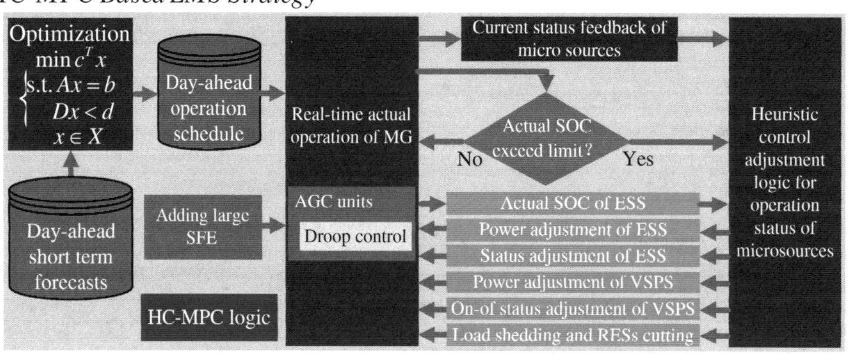

Figure 2. HC-MPC based EMS strategy.

Rolling optimization is poor at handling inputs with large SFE whereas heuristic control is good at coping with large SFE despite that the operation is usually not optimal. Day-ahead schedule is utilized as command reference, operation economy can be improved. Real-time operating status of VSPS and ESS and feedback of SOC are taken by the online heuristic controller to determine the on-off status of VSPS and decide the power of VSPS and ESS, as shown in figure 2.

3. Case Study

3.1. Simulation Setup

Hybrid-MPC is tested in self-developed program named MG real operation simulation (MG-ROS). In MG-ROS droop control is adopted by VSPS and ESS. Schedule commands are taken as droop curves power set points. Day-ahead forecast for net load plus SFE are taken as real-time actual inputs of MG.

3.2. Day-ahead Operation and Operation with Small SFE

In the first case, no SFE is considered and day-ahead forecast of net load is regarded as ideal input for MG. Optimal optimization result is shown in figure 3(b). It is easy to find that SOC is always within the normal range which indicates that status of ESS is normal throughout the whole operation periods.

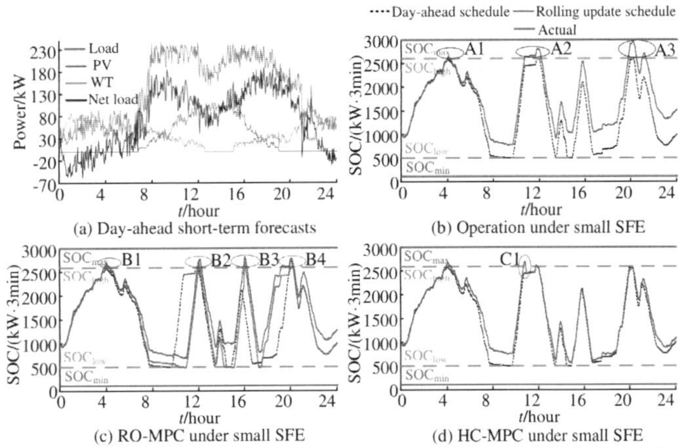

Figure 3. Day-ahead short-term forecasts and SOC of ESS.

Then small SFE of net load is added to the operation of MG and three cases are carried out where no EMS strategy, RO-MPC and HC-MPC is applied respectively to observe the impacts of SFE on SOC of ESS. To obtain sample of SFE, data within the error 95% confidence interval is taken as small SFE, and data within the interval between error 95% and 98% confidence intervals is taken as large SFE.

It is clearly observable in figure 3(b) that the difference between schedule SOC and actual SOC is increasing gradually and at A3 actual SOC exceeds the normal range which indicates that overcharge of ESS has happened. In figure 3(c) the operation schedule of MG has changed four times from B1 to B4 and the effect of RO-MPC is obvious. SOC is adjusted and kept within extreme operation limit. In figure 3(d) the operation status of micro sources is adjusted and actual SOC is modified within several consecutive periods around C1. By using HC-MPC, deviation of SOC is successfully reduced.

3.3. Operation with Large and Mixed SFE

Large and mixed SFE of net load is added to the operation of MG and four cases are carried out where no strategy, RO-MPC, HC-MPC and Hybrid-MPC is applied respectively.

It can be seen in figure 4(a) that deviation of SOC increases rapidly and capacity of ESS must be very large to keep the stable operation. It is clear in figure 4(b) that applying RO-MPC the capacity of ESS must increase to 4000. It is observable in figure 4(c) that HC-MPC can successfully maintain actual SOC within normal range. It is unnecessary to increase the capacity of ESS. However, there is a significant increase in amount of load shedding and RESs cutting. In figure 4(d) Hybrid-MPC can successfully maintain the actual SOC within normal range and there is no need to increase the capacity of ESS. The total amount of load shedding and RESs cutting is significantly less. The comparison of additional simulation results are shown in figure 5 and figure 6.

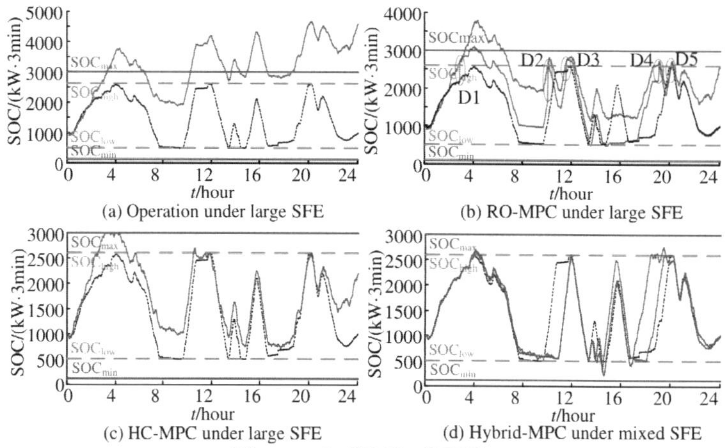

Figure 4. SOC of ESS.

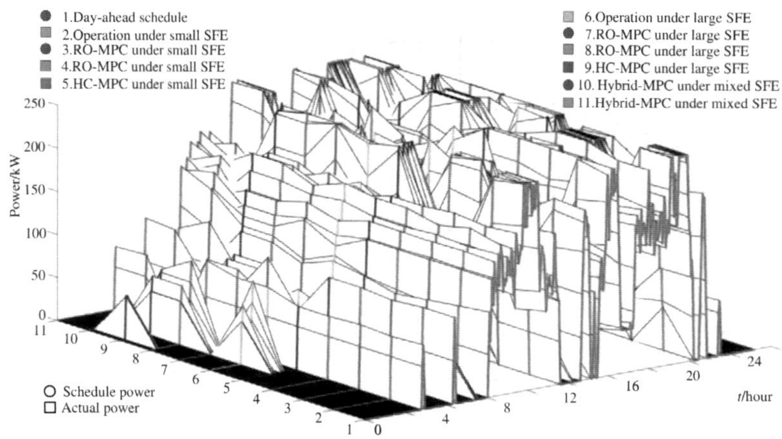

Figure 5. Comparison of power of VSPS.

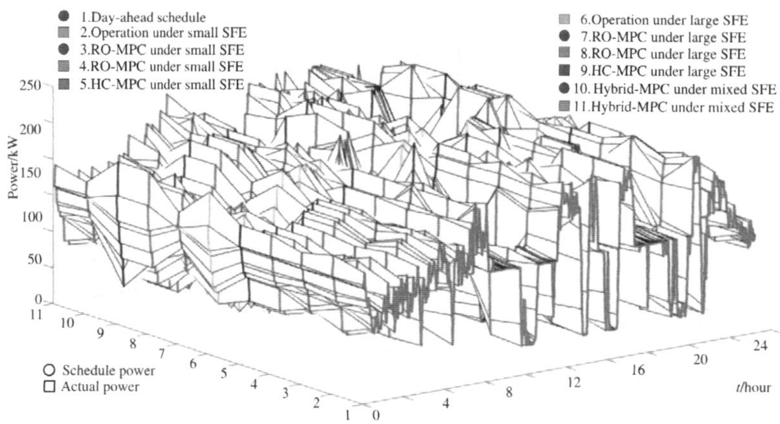

Figure 6. Comparison of power of ESS.

4. Conclusion

This paper has presented a novel online Hybrid-MPC based EMS strategy to cope with SOC deviation of ESS. Hybrid-MPC consists of decreasing-horizon rolling optimization, heuristic cooperative control, and closed-loop feedback correction. Decreasing-horizon rolling optimization is utilized to handle with small disturbances, while heuristic control is applied to cope with large disturbances. Feedback correction is used to eliminate SOC deviation. The effectiveness of the strategy has been tested in MG-ROS. Results demonstrated that the strategy has good performance on handling disturbances, keeping SOC within normal range and preventing the unstable operation of ESS.

5. References

[1] D. E. Olivares, C. A. Cañizares and M. Kazerani, "A Centralized Energy Management System for Isolated Microgrids," in IEEE Transactions on Smart Grid, vol. 5, no. 4, pp. 1864-1875, July 2014.
[2] L. Zhou, F. Li and X. Tong, "Active Network Management Considering Wind and Load Forecasting Error," in IEEE Transactions on Smart Grid, vol. PP, no. 99, pp. 1-8.
[3] H. Bludszuweit, J. A. Dominguez-Navarro and A. Llombart, "Statistical Analysis of Wind Power Forecast Error," in IEEE Transactions on Power Systems, vol. 23, no. 3, pp. 983-991, Aug. 2008.
[4] Yahong Chen, Changhong Deng, et al, Impacts of stochastic forecast errors of renewable energy generation and load demands on microgrid operation, Renewable Energy, Volume 133, 2019, Pages 442-461.
[5] J. Wang and M. Shahidehpour, "Security-Constrained Unit Commitment With Volatile Wind Power Generation," in IEEE Transactions on Power Systems, vol. 23, no. 3, pp. 1319-1327, Aug. 2008.
[6] M. Carrion and J. M. Arroyo, "A computationally efficient mixed-integer linear formulation for the thermal unit commitment problem," in IEEE Transactions on Power Systems, vol. 21, no. 3, pp. 1371-1378, Aug. 2006.
[7] Alessandra Parisio, Evangelos Rikos and George Tzamalis, "Use of model predictive control for experimental microgrid optimization," Applied Energy, Volume 115, 2014, Pages 37-46.
[8] Hong Wang, Yanjun Huang, Amir Khajepour, Qiang Song, "Model predictive control-based energy management strategy for a series hybrid electric tracked vehicle," Applied Energy, Volume 182, 2016, 105-114.
[9] N. Liu et al., "A Heuristic Operation Strategy for Commercial Building Microgrids Containing EVs and PV System," in IEEE Transactions on Industrial Electronics, vol. 62, no. 4, pp. 2560-2570, April 2015.
[10] W. Su, J. Wang and J. Roh, "Stochastic Energy Scheduling in Microgrids With Intermittent Renewable Energy Resources," in IEEE Transactions on Smart Grid, vol. 5, no. 4, pp. 1876-1883, July 2014.
[11] R. Palma-Behnke et al., "A Microgrid Energy Management System Based on the Rolling Horizon Strategy," in IEEE Transactions on Smart Grid, vol.

Acknowledgments

This work was supported by The National Key Research and Development Program of China (2017YFB0903705).

IOP Conf. Series: Earth and Environmental Science **342** (2019) 012017 doi:10.1088/1755-1315/342/1/012017

Study on doubly closed loop control strategy and simulation analyzing of pmsg wind turbine with back to back converter

CY Gong[1], JL Zou[2]

[1] School of Automation Engineering, Shanghai University of Electric Power, 2588 Changyang Road, Shanghai 200090, China
[2] Jiaxing Renewable Power Electrical Co., Ltd, 1888 West Zhongshan Road, Jiaxing Zhejiang 314031, China

Email: 18508528566@163.com

Abstract. The generation technology of wind turbine connected to grid is a hot research in renewable energy application field. Based on doubly closed loop control strategy for the converter, the paper designed the mathematical model and research the attributes of PMSG wind power generation system, adopted double-loop control strategy for generator and grid converter, which is the corresponding main basis to build the entire simulation modelling step by step with Simulink, to control the active and reactive power effectively. Moreover, current dynamic response is fast apparently and the system has capacity to recover from a disturbance. The feasibility and performance of this model had been verified by the Matlab/Simulink, showen the strength and robustness of PMSG system from a disturbance circumstance.

1. Introduction

Nowadays, facing to the increasingly serious energy crisis and the further deterioration of supply-demand, optimize the energy structure and develop the clean and environmentally friendly renewable energy is becoming imminent[1].Compared to the CSCF (constant speed constant frequency) wind turbine generator, the VSCF (variable speed constant frequency) could achieve higher utilization of wind power, which has been occupying the manufacturing industries and markets of wind power generation gradually[2]. Wherein, DFIG (Doubly-Fed Induction Generator) and PMSG (Permanent Magnet Synchronous Generator) are adopted widely by researchers as the typical VSCF. Owing to the irreplaceable need of gearbox, worse LVRT (Low Voltage Running Through) ability and could not be decoupled to control the converter independently without rotor position sensor, the PMSG would be considered to establish the framework of simulation modeling eventually[3, 4].

The core component of whole modeling is a pair of back to back converter, which equipped with the fundamental double-loop control strategy. To ensure the voltage on DC bus at a stable value and make the output power of stator reach the decoupled level totally, the control strategy should be distinguished into the generator side as well as the grid side. This control strategy is based on the VOC (vector oriented control) method, which would decouple the AC side current into active and reactive current component through the rotated coordinate transformation, thereby could generate the inner current loop and outer DC voltage loop[5]. The double-loop control strategy could enhance the precision and accuracy of control and the ability to inhibit the load disturbance due to add the inner current loop using the characteristics for fast response of current[6].

2. The mathematical model of PMSG wind power generation system

The typical framework of PMSG wind power generation system is consists of a wind turbine, a PMSG, a pair of back to back converter and the grid. In theory, the value of wind turbine could transform wind power into the electrical power correspond the formula (the Betz Theorem). And for the variable speed wind turbine, there is a relation between angular velocity of wind turbine and wind speed in Equation (1):

$$\lambda = \frac{R\omega}{\upsilon} \tag{1}$$

where λ is the tip ratio. The variable speed wind turbine can keep operating in the most efficient situation (the C_p reaches the maximum) through regulating the λ at different wind speed, which could achieve the maximum power point tracking (MPPT) [7]. The primary designed principle is according to the Newton's second law as follows:

$$T_m - T_e = H\frac{d\omega_m}{dt} \tag{2}$$

where T_m is the mechanical torque, T_e is the electromagnetic torque, ω_m is the angular velocity of generator and H is the rotated inertia.

Due to the power is equal to torque multiplies angular velocity and the θ is the differential coefficient of angular velocity on the basis of foundations of physics, the mathematical model of PMSG would be designed as Equation (3), which is equivalent to a DC generator could be controlled independently, where the Φ is the flux regard as a constant at here.

$$\begin{cases} u = E\sin(\omega t + \theta) \\ E = k\Phi\omega_m \end{cases} \tag{3}$$

3. The control strategy of converter

The control design of the grid side converter is to ensure the power from generator side could be controlled, which makes the DC-link and AC voltage constant. Hence, establish double closed loop according to the corresponding objectives. The outer voltage loop is planned to control the DC-link and AC voltage to achieve and keep the demanded value, while the inner current loop would regulate the power delivered.

It is necessary to mention that proportion and integration parameter of current loop would be about five times greater than the voltage loop, so the response speed of inner loop will be faster than the outer loop, which means the reference value of stator current on the direct axis could be decided. The control structure diagram of grid side converter had been showed in Figure 1 (a). Compare with the grid side converter, the main objective of the generator side is to totally decouple the power from the wind turbine transformed and could control it independently. Besides the outer voltage loop combined with inner current loop as well, the maximum power point tracking decided the reference value of stator current on the direct axis. The control structure diagram of generator side converter had been showed in Figure 1 (b).

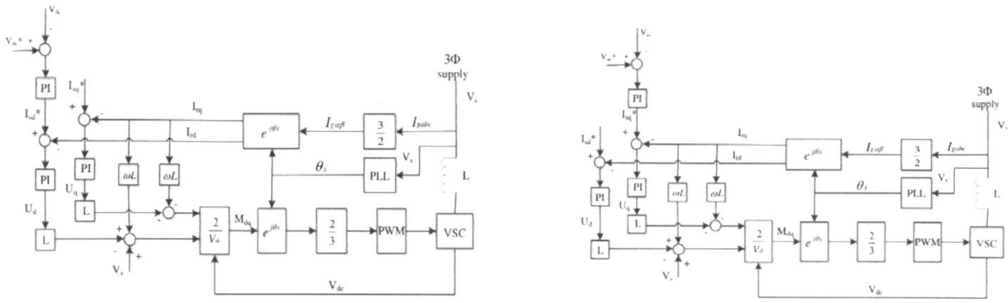

(a)grid side (b) generator side
Figure 1 Control structure diagram of converter.

4. Simulation of PMSG wind power generation system

4.1. Design of modeling with Simulink

The sketch of the whole modeling simulation is shown as Figure2. A wind turbine block connected to a pair of back-back converters which the state would be controlled by two controllers at the bottom of this diagram. For both of generator and grid side controllers, there are phase locked loop (PLL) and DQ transformation to achieve the control objectives apart from the double closed loop[8]. Besides the phase reactance and line impedance existing in the conventional circuit as well as some measurement blocks used for the AC or DC voltage and current, the circuit also included the 3-phase circuit breaker, which could causes the disturbance or fault depending on the value of resistance in operating process and the auxiliary circuit could help system remain steady state when the disturbance appears at AC side. Set 3MW for the rated capacity of wind turbine and the ratio of transformer is 690/11000 (V), additionally, the phase reactance had been decided to 0.1587Ω and inductance would be 0.10104mH at the rated frequency 50Hz in the case of considering 20% only. The sampling adopts the discrete method and the period is 2μs.

The total built process of modeling would be divided into four steps. The initial modeling consists of a DC voltage source (battery), single converter with controller, a transformer and an AC voltage source network. Then replace the DC voltage source with a large capacitor which is set 50mF and determine the reference voltage on the capacitor is 1300V. In the next step, copy the VSC and its controller except the voltage loop and combine with another AC voltage source, which has been put on one side of circuit connecting the capacitor. Finally, replace the AC voltage source network using the wind turbine block at the input side, which is the final structure of simulation modeling like the Figure 2.

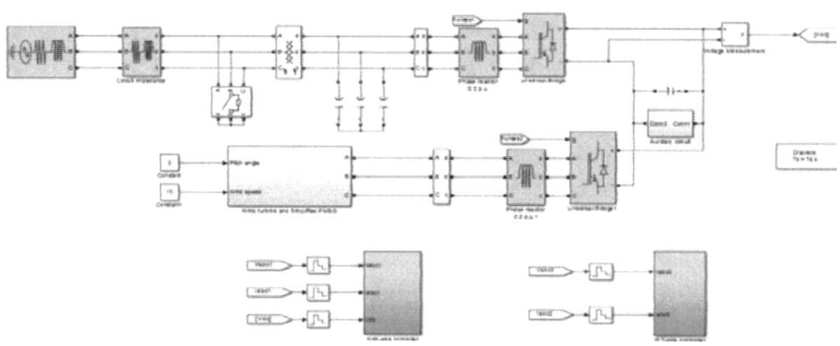

Figure 2 Simulation modeling of PMSG wind power generation system.

4.2. Analysis of simulation results

Try to run this modeling under the circumstances that set a step signal for wind speed which the initial magnitude is 8m/s and final magnitude is 10m/s at 0.6ms, moreover, according to the short circuit ratio, set 10 Ω as the magnitude of single phase resistance in three phase circuit breaker, which would turn on at 1.0ms and turn off at 1.1ms. The outcomes are shown in Figure 3 and Figure 4.

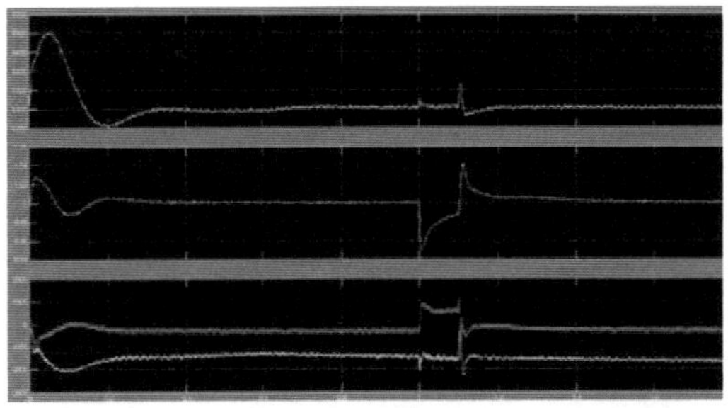

Figure 3. Waveforms of DC-link voltage, AC voltage, i_d and i_q.

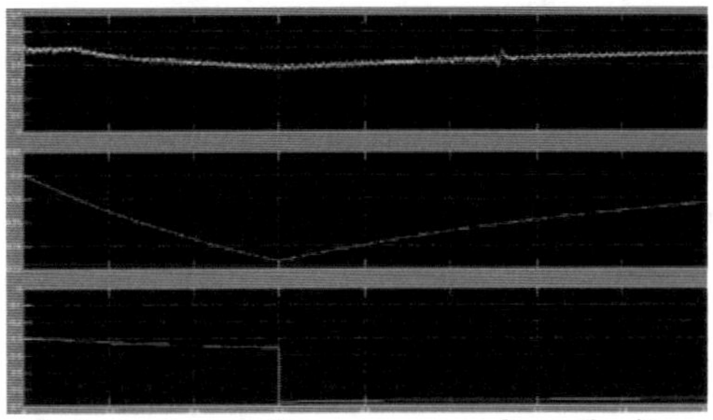

Figure 4. Waveforms of electromagnetic power angular velocity and mechanical torque.

Basically, the direct and quadrate current on the stator would track required value immediately, which means that it has good dynamic response attribute. And the DC-link voltage could also track and remain the required value, which means the output power could be controlled successfully. In whole simulation process, initially, because the wind speed is not large enough, the input power could not provide the required DC-link voltage before 0.6ms. Furthermore, it is obvious to see that the DC-link and AC voltage has returned to the required value after 1.0ms, which means the system is so strong that could recover from a disturbance. Besides, the angular velocity is large at beginning, which causes the tip ratio is large too, so the power coefficient would be extracted on the right side of the maximum point and get close it gradually, which could increase the power that wind turbine transformed and the mechanical torque, while the electromagnetic power would show the declining tendency. However, when the wind speed increases, the power coefficient would start to get away from the maximum point. Hence, all parameters would present the opposite tendency compared with the previous condition absolutely.

5. Conclusion

For the PMSG wind power generation system, this paper purposed and designed a double closed loop control strategy on the basis of PWM technology with converter, which could achieve the objective to control the output power from wind turbine. When a disturbance had been appearing from outer environment, designed system could be able to recover to normal operating situation. At the same time, the dynamic response speed had been still fast. The entire modeling system had been ensured that it could be adopted in laboratory research after simulation with Matlab. Finally, it is necessary to prove that this system could recover from a more serious disturbance or a fault operating circumstance, which is the main shortage that needs to be completed in the future plan.

Acknowledgment

This work was financially supported by the National 863 Program under Grant 2014AA052005.

Reference

[1] A. Zahedi, "Current status and future prospects of the wind energy," 2012 10th International Power & Energy Conference (IPEC), Ho Chi Minh City, 2012, pp. 54-58.

[2] Feng Gao, Ning An, Ning-Hui Zhu and Feng Su, "Comparison of operation characteristics of variable speed constant frequency wind farms into power system," 2011 IEEE Power Engineering and Automation Conference, Wuhan, 2011, pp. 170-174.

[3] R. Datta and V. T. Ranganathan, "Variable-speed wind power generation using doubly fed wound rotor induction machine-a comparison with alternative schemes," in IEEE Transactions on Energy Conversion, vol. 17, no. 3, pp. 414-421, Sept. 2002.

[4] K. Patil and B. Mehta, "Modeling and simulation of variable speed wind turbine with direct drive permanent magnet synchronous generator," 2014 International Conference on Green Computing Communication and Electrical Engineering (ICGCCEE), Coimbatore, 2014, pp. 1-6.

[5] N. K. Jena, K. B. Mohanty, H. Pradhan and S. K. Sanyal, "A decoupled control strategy for a grid connected direct-drive PMSG based variable speed wind turbine system," 2015 International Conference on Energy, Power and Environment: Towards Sustainable Growth (ICEPE), Shillong, 2015, pp. 1-6.

[6] M. Hannachi and M. Benhamed, "Modeling and control of a variable speed wind turbine with a permanent magnet synchronous generator," 2017 International Conference on Green Energy Conversion Systems (GECS), Hammamet, 2017, pp. 1-6.

[7] Yiguang Chen, Zhiqiang Wang, Yonghuan Shen and Lingbing Kong, "A control strategy of direct driven permanent magnet synchronous generator for maximum power point tracking in wind turbine application," 2008 International Conference on Electrical Machines and Systems, Wuhan, 2008, pp. 3921-3926.

[8] P. Wisartpong, J. Koseeyaporn and P. Wardkein, "Pulse width modulation based on phase locked loop," 2008 5th International Conference on Electrical Engineering/Electronics, Computer, Telecommunications and Information Technology, Krabi, 2008, pp. 697-700.

Research on co-integration and causal relationship between China's energy consumption and economic growth

Y S Huang, M S Shi, Y Q Yang

Department of Economics and Management, North China Electric Power University, Hebei, 071003, China

Email: shimengshu1227@foxmail.com

Abstract. This paper selects the annual data of China's total energy consumption and GDP from 1978 to 2016 as a research sample, and uses Eviews to perform unit root test, cointegration test, Engle-Grange causality test and VEC model on GDP and energy consumption time series. The results show that the first-order difference between GDP and total energy consumption is stable at a significant level of 10%; GDP, total energy consumption and sub-category energy consumption have a co-integration relationship at a significant level of 5%; Engle-Grange causality test results show that at the 5% significant level, the causal relationship between GDP and energy consumption is obvious, and there is a two-way Granger causality.

1. Introduction

The annual report of World Energy Outlook released by the International Energy Agency (IEA) in September 2010 predicts that China's energy demand will surge by 75% between 2008 and 2035, and by 2035 China's energy demand will grow twice as fast as the world. This paper explores the relationship between China's economic growth and energy consumption, and takes 2006 as a node to analyze the changes of the relationship between energy consumption and economic growth before and after China put forward the "energy saving and emission reduction policy".

Chang et al. (2013) applied panel causality test to examine the relationship between economic growth and energy consumption in 12 Asian countries from 1970 to 2010, and concluded that the two are causal [1]. Jalil Abdul & Feridun Mete (2014) used ARDL model to discuss the relationship between economic growth and energy consumption in China from 1952 to 2008. The results show that energy consumption plays a more significant role in promoting China's economic growth [2]. Based on the data from 1990 to 2011, Liu Xiantao and Shijun (2014) used co-integration analysis to study the long-term relationship between energy consumption and economic growth in China. The results showed that there was no stable long-term relationship between them [3]. Industrial structure, technological progress, population, economic growth and so on affect the energy efficiency of each province, and have a long-term balanced and stable relationship [4]. In the short run, a big driving force of China's economic development is the consumption of new energy, but in the long run, the development of economy in turn promotes the development of new energy [5].

2. Empirical analysis

2.1. Data processing

(1) Data sources

From 1978 to 2016, China's total energy consumption, oil consumption, coal consumption and electricity consumption came from the National Bureau of Statistics and the National Energy Administration of China. 21978-2016 China's population and GDP data are from the National Bureau of Statistics.

(2) Symbolic representation

Table 1. Symbolic Representation.

Symbol	Representative content
GDP	Per capita GDP (10,000 yuan per person)
PGC	Per capita natural gas consumption (tons of standard coal/person)
TEC	Total energy consumption per capita (tons of standard coal/person)
POC	Per capita oil consumption (tonnes of standard coal/person)
PCC	Per capita coal consumption (tons of standard coal/person)
PEC	Per capita electricity consumption (tonnes of standard coal/person))

(3) Data processing

In this study, China's per capita energy consumption and per capita GDP data are used to eliminate the impact of population growth; since the data logarithm does not change the original co-integration relationship, and can linearize its trend, eliminate the heteroscedasticity in the time series, this paper carries out natural logarithm transformation of variables. Because the price factor has a great influence on GDP, the GDP data are processed in the form of 1978 as the base year.

Through the original data graph in figure 1, we can find that the growth rate of total energy consumption and total coal consumption is obviously faster than the growth trend of economic growth. The development trend of different energy consumption is more synchronous.

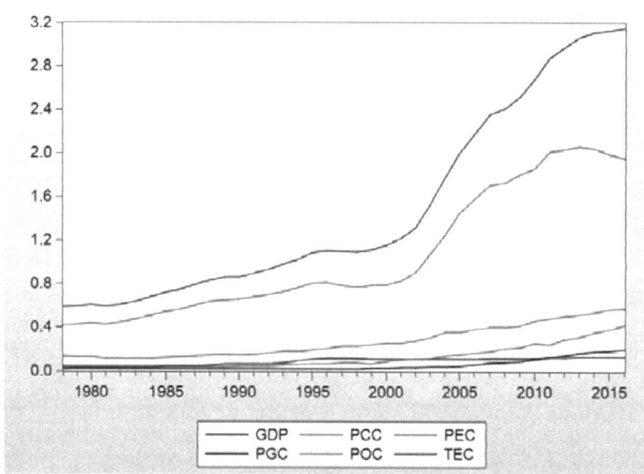

Figure 1. Economic growth and energy consumption trend chart 1978-2016.

2.2. The stability test of sequences

The unit root test of GDP and energy consumption series is to avoid two variables of non-stationary time series. In the subsequent cointegration and causality test, the original two unrelated variables appear the fallacy of causality or the conclusion of meaningless regression.

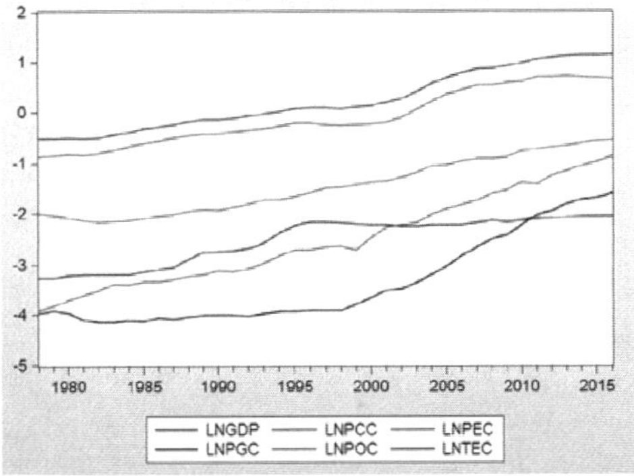

Figure 2. Trends in Economic Growth and Energy Consumption (Logarithmic Processing) 1978-2016.

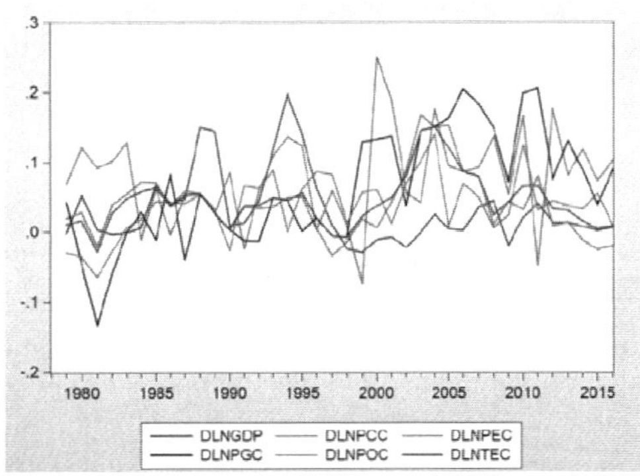

Figure 3. First-order differential trend chart of variables 1978-2016.

Through the graphical analysis of the variables, it can be found in figure 2 that all the six variables have the same trend of growth, the direction of change is basically the same, and the variables show an unstable upward trend. The ADF test of the variables is further carried out.

It can be seen from the figure 3 that the first-order difference of each variable fluctuates near zero and tends to be stable. Next, the unit root test is performed on the variables and the results are performed in table 2.

Table 2. Unit Root Test Results.

Variables	Test type（c,t,k）	ADF	P	conclusion
lngdp	2	-1.0091	0.9301	Unstable
dlngdp	1	-3.5933	0.0046	stable
lntec	1	-2.8956	0.1754	Unstable
dlntec	1	-2.6362	0.0953	stable
lnpcc	1	-2.8798	0.1803	Unstable
dlnpcc	0	-1.7551	0.0753	stable
lnpec	0	-1.3143	0.8691	Unstable
dlnpec	0	-6.7468	0.0000	stable
lnpoc	0	-4.5026	0.0048	stable
dlnpoc	0	-4.6123	0.0037	stable
lnpgc	1	-2.0681	0.5459	Unstable
dlnpgc	0	-4.1636	0.0117	stable

c,t,k 123 represents intercept term, trend term and lag order respectively. The selection criteria of lag period refer to AIC and SC criteria.

Through unit root test of variables in table 2, we can find that per capita GDP, per capita total energy consumption, per capita electricity consumption, per capita coal consumption and per capita natural gas consumption are first-order stable at 10% confidence level, and their mean and variance do not change with time, showing the same-order monolithic. The co-integration relationship between variables can be further validated. However, the original sequence of per capita oil consumption is stable, which is not the same order as GDP, so no further test is made.

2.3. Establishing VAR model to determine lag time

Table 3. Variable Delay Period.

Lag	LogL	LR	FPE	AIC	SC	HQ
0	137.6259	NA	4.34e-10	-7.368104	-7.148171	-7.291341
1	392.3130	424.4786	1.27e-15	-20.12850	-18.80890*	-19.66793
2	435.4460	59.90692*	5.04e-16*	-21.13589*	-18.71662	-20.29150*
3	457.5946	24.60951	7.45e-16	-20.97748	-17.45855	-19.74928

A VAR model is established for stationary variables of the same order to determine their lag time. It can be found from the table 3 that according to AIC criterion and considering the comprehensive test, the second-order lag period is adopted for variables.

2.4. Cointegration test

Table 4. Cointegration Test Results.

Hypothesized No. of CE(s)	Eigenvalue	Trace Statistic	0.05 Critical Value	Prob.**
None *	0.798872	102.2310	69.81889	0.0000
At most 1	0.417376	42.88990	47.85613	0.1353
At most 2	0.318838	22.90200	29.79707	0.2509
At most 3	0.209295	8.695642	15.49471	0.3943
At most 4	0.000187	0.006904	3.841466	0.9332
Trace test indicates 1 cointegration at the 0.05 level				
None *	0.798872	59.34113	33.87687	0.0000
At most 1	0.417376	19.98790	27.58434	0.3420
At most 2	0.318838	14.20636	21.13162	0.3483
At most 3	0.209295	8.688738	14.26460	0.3130
At most 4	0.000187	0.006904	3.841466	0.9332
Max-eigenvalue test indicates1 cointegration at the 0.05 level				

Johansen cointegration test is performed in the order of the number of cointegration relations r=0 to r=k-1 until the corresponding original hypothesis is rejected. Table 4 shows the results of cointegration test. In order to illustrate the test results, the first column of "None" means that the original hypothesis of the test is "there is zero co-integration relationship", under which the trace statistics are equal to 102.2310, the critical value of 5% is equal to 69.81899, and the trace statistics are greater than the critical value. Therefore, the original hypothesis is rejected, indicating that there is at least one co-integration relationship. Then we examine the original hypothesis of "Atmost1", which indicates that there is at most one co-integration relationship. The trace statistics under this hypothesis are 42.88990, and the critical value is 47.85613, which is less than 5%. Therefore, we accept the original hypothesis and show that there is at least one co-integration relationship. Trace statistics test results show that there is a co-integration relationship at the level of 5%.

The results of maximum eigenvalue test are not consistent with those of trace statistics test. Trace statistics are generally the main test because they are more effective in general. The test results of trace statistics are consistent with the test results of maximum eigenvalue. Therefore, there is a co-integration relationship between per capita GDP, per capita energy consumption, per capita electricity consumption, per capita coal consumption and per capita natural gas consumption in China from 1978 to 2016. Cointegration equation 7 is:

$$LNGDP = 11.59286LNTEC - 7.783920LNPCC - 1.876573LNPEC - 0.749156LNPGC \tag{7}$$

Cointegration equation shows that LNGDP and LNTEC have a positive long-term equilibrium relationship, that is, every 1% increase in total energy consumption leads to an 11.59% increase in GDP, while GDP and coal consumption, electricity consumption and natural gas consumption show a reverse long-term equilibrium relationship, because the total energy consumption is composed of various types of energy consumption, energy consumption and economic growth. The co-integration relationship is also in line with China's actual situation. The overall scale of economic growth will have an impact on energy consumption. With the continuous expansion of economic scale, energy consumption will inevitably increase.

2.5. Establishment of VEC model

2.5.1. Model Estimation. The long-term equilibrium relationship between energy consumption and economic growth can be reflected by the co-integration equation. The short-term change of energy consumption can be analyzed by using VEC model. Table 5 shows the parameter estimation results of VEC Model.

Table 5. Parameter Estimation Results of VEC Model.

Error Correction:	D(LNGDP)	D(LNPCC)	D(LNPEC)	D(LNPGC)	D(LNTEC)
CointEq1	-0.002279	-0.009734	0.012745	-0.054189	-0.014174
D(LNGDP(-1))	0.826296	-0.147886	0.086792	0.053819	-0.107152
D(LNGDP(-2))	-0.310306	-0.100208	-0.074743	-0.317751	-0.047824
D(LNPCC(-1))	-0.355027	0.771373	1.274628	1.915042	0.921596
D(LNPCC(-2))	-0.094231	0.436637	-0.259216	1.090312	0.343050
D(LNPEC(-1))	-0.124262	0.150046	-0.050026	-0.090998	0.062181
D(LNPEC(-2))	-0.098494	0.040149	-0.143535	-0.357333	-0.012822
D(LNPGC(-1))	-0.166303	-0.101410	0.441273	0.181049	-0.046167
D(LNPGC(-2))	-0.033412	-0.137134	0.272381	0.027518	-0.085343
D(LNTEC(-1))	0.853897	-0.031968	-1.037935	-1.964717	-0.332858
D(LNTEC(-2))	-0.006164	-0.459417	-0.654482	-1.457387	-0.531334
C	0.025400	0.019827	0.084351	0.124658	0.040300
R-squared	0.576415	0.689618	0.299358	0.815062	0.749700
Adj.R-squared	0.382272	0.547360	-0.021769	0.730299	0.634979

$$Co\text{int}\,Eq1 = LNGDP - 12.82490LNPCC - 9.083319LNPEC + 2.689285LNPGC + 18.88082LNTEC - 16.68210 \quad (8)$$

The expression of the error correction term is consistent with that of Johansen's cointegration test, but the coefficients are estimated differently by introducing a constant term into the cointegration test relation. The expression of VEC model is as follows:

$$\Delta Y_{t-1} = \begin{bmatrix} -0.002 \\ -0.010 \\ 0.013 \\ -0.054 \\ -0.014 \end{bmatrix} Co\text{int}\,EQ_{t-1} + \begin{bmatrix} 0.83 & -0.36 & -0.12 & -0.16 & 0.85 \\ -0.15 & 0.77 & 0.15 & -0.10 & -0.03 \\ 0.09 & 1.27 & -0.05 & 0.44 & -1.04 \\ 0.05 & 1.92 & -0.09 & 0.18 & -1.96 \\ -0.11 & 0.92 & 0.06 & -0.05 & -0.33 \end{bmatrix} \Delta Y_{t-1} + \begin{bmatrix} -0.31 & -0.09 & -0.10 & -0.03 & -0.01 \\ -0.10 & 0.44 & 0.04 & -0.14 & -0.46 \\ -0.07 & -0.26 & -0.14 & 0.27 & -0.65 \\ -0.32 & 1.09 & -0.36 & 0.03 & -1.46 \\ -0.05 & 0.34 & -0.01 & -0.09 & -0.53 \end{bmatrix} + \varepsilon_t; \quad (9)$$

$$\Delta Y = \begin{bmatrix} D(LNGDP) \\ D(LNPCC) \\ D(LNPEC) \\ D(LNPGC) \\ D(LNTEC) \end{bmatrix} \quad (10)$$

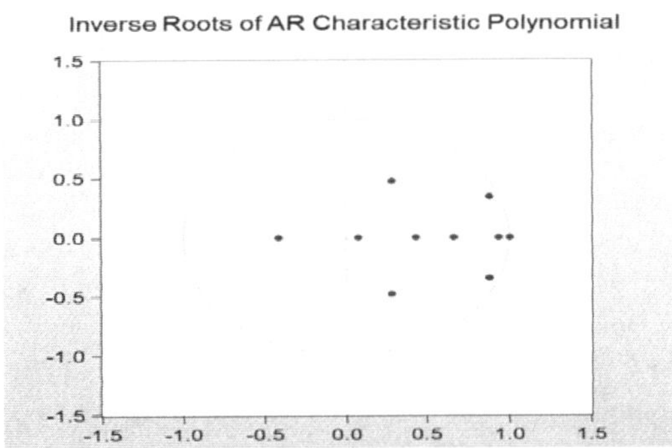

Figure 4. VEC stationarity test.

For the VAR model with K endogenous variables and P * k roots, the AR characteristic polynomial has 6 roots. If the reciprocal of all roots of the estimated VAR model is less than 1, i.e. within the unit circle, the VAR model is stable. The figure 4 shows that the 10 roots involved in the VEC model are in the unit circle, which shows that the VEC model is stable.

IOP Conf. Series: Earth and Environmental Science **342** (2019) 012018 doi:10.1088/1755-1315/342/1/012018

2.5.2. Impulse response function

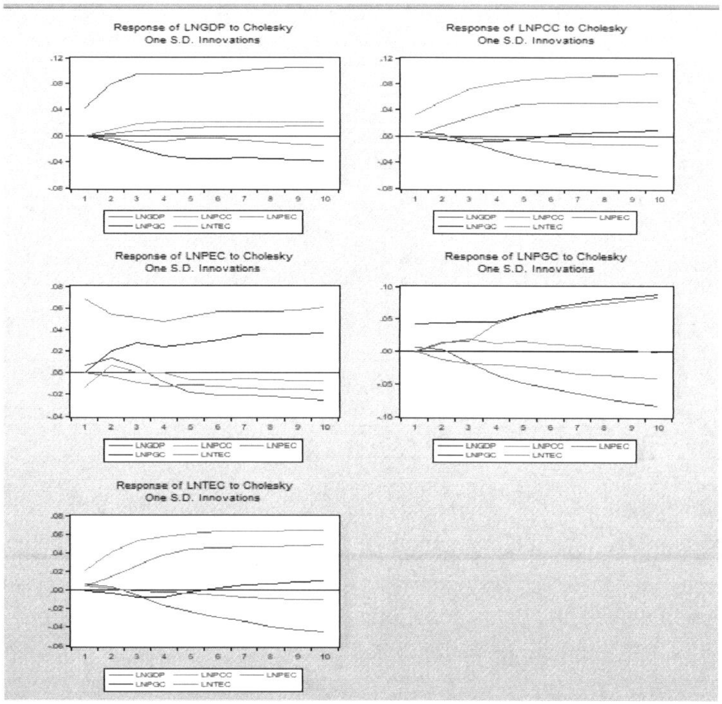

Figure 5. Diagram of impulse response function.

Through the impulse response function in figure 5, we can find the impact of a standard deviation of energy consumption on GDP. After being shocked by energy consumption, GDP began to rise in the first period and maintained a stable trend in the fourth period.

2.5.3. Variance decomposition

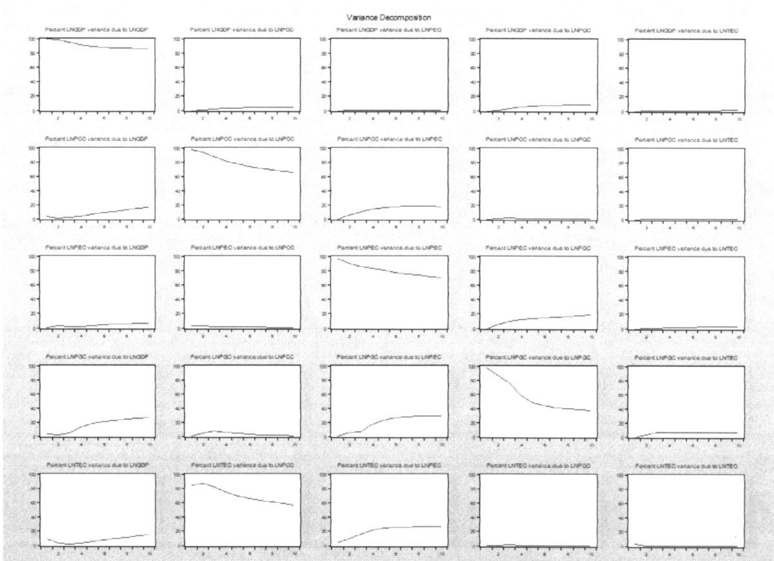

Figure 6. Variance Decomposition Diagram.

From the variance decomposition graph in figure 6, it can be found that the explanatory part of variables decreases gradually with the increase of periods, while the explanatory part of other variables increases gradually. The figure shows that the explanatory part of total energy consumption to GDP changes increases gradually and reaches the peak in the tenth period, that is, about 20% of the variance of GDP changes can be explained by the change of total energy consumption.

2.6. Granger causality test

Table 6. Granger causality test results.

Hypothesis	Lagged term	F	P	conclusion
LNPCC is not the Granger Cause of LNGDP	2	0.96152	0.3931	No rejection
LNGDP is not the Granger Cause of LNPCC	2	3.39579	0.0460	rejection
LNPEC is not the Granger Cause of LNGDP	2	0.28239	0.7558	No rejection
LNGDP is not the Granger Cause of LNPEC	2	0.06302	0.9390	No rejection
LNPGC is not the Granger Cause of LNGDP	2	0.35770	0.7020	No rejection
LNGDP is not the Granger Cause of LNPGC	2	4.47117	0.0194	rejection
LNTEC is not the Granger Cause of LNGDP	2	1.27008	0.2946	No rejection
LNGDP is not the Granger Cause of LNTEC	2	5.99163	0.0062	rejection
LNTEC is not the Granger Cause of LNPCC	2	3.82641	0.0324	rejection
LNPCC is not the Granger Cause of LNTEC	2	1.76166	0.1880	No rejection
LNPGC is not the Granger Cause of LNPEC	2	2.09412	0.1397	No rejection
LNPEC is not the Granger Cause of LNPGC	2	7.09636	0.0028	rejection
LNTEC is not the Granger Cause of LNPEC	2	0.88014	0.4245	No rejection
LNPEC is not the Granger Cause of LNTEC	2	5.70215	0.0076	rejection
LNTEC is not the Granger Cause of LNPGC	2	10.5824	0.0003	rejection
LNPGC is not the Granger Cause of LNTEC	2	0.18410	0.8327	No rejection

Granger causality in table 6 shows that there is a single Granger causality between coal consumption and GDP in the 5% confidence interval, while there is no Granger causality between electricity consumption and GDP, there is a single causality between natural gas consumption and GDP, and there is a two-way Granger causality between total energy consumption and GDP. That is to say, the growth of economy leads to the increase of energy consumption, and the increase of energy consumption promotes economic growth. This is because with the rapid economic growth, the increase of carbon-based energy consumption will inevitably accompany the increase of energy consumption. Economic growth is the main driving factor of the increase of energy consumption. With the increase of energy consumption, the economy will also grow. This is because China's economic growth is accumulated by energy consumption, which also means China's extensive economic growth mode.

3. Conclusion

The relationship between China's economic growth and energy consumption shows that China's economic growth is inseparable from energy consumption to a certain extent, and energy consumption brings a series of environmental pollution problems, coupled with the increasing depletion of non-renewable energy, resulting in resource-dependent development difficult to sustain.

As China is in the process of economic development, it is inevitable to experience some major events that have had a huge impact on the Chinese economy. The oil crisis of 1973; reform and opening up policy was proposed in 1978; the financial earthquake in 1987; the establishment of the socialist market economic system in 1992, the exploration stage of the market economy; the Asian financial crisis in 1997; the global financial crisis in 2008; 2016 is in a stable phase of the market economy. These events have a certain degree of influence on the relationship between China's economy and energy, which will cause certain errors in the long-term cointegration and causal relationship analysis of China's energy and economy. In the future, the research will go deep into the time nodes of major events and further talk about the development relationship between China's energy and economy.

Acknowledgments

This work was supported by the Energy Economic Development Strategy Research project of the Science and Technology Department of Hebei Province (grant number 12457206D-22).

IOP Conf. Series: Earth and Environmental Science **342** (2019) 012018 doi:10.1088/1755-1315/342/1/012018

Reference

[1] Chang Tsangyao, Chu Hsiao-Ping and Chen Wen-Yi. Energy consumption and economic growth in 12 Asian countries: panel data analysis. 2013 *Applied Eco-nomics Letters* **20** 282-87

[2] Jalil Abdul and Feridun Mete. Energy-Driven economic growth: energy consumption-economic growth nexus revisited for China. 2014 *Emerging Markets Finance &amp Trade* **50** 159-68

[3] Liu X T and Shi J. Vector autoregressive model test analysis of energy consumption and economic growth in China. 2014 *Statistics and decision-making*

[4] Zhou R and Wang X H. Empirical analysis of energy utilization and regional economic development, industrial structure. 2017 *Statistics and Decision-making* 139-42

[5] Guo F G, Chen G and Du N S. Empirical analysis of the relationship between China's new energy consumption and economic growth. 2012 *Enterprise Economy* **31** 35-37

Prospect of Natural Gas Distribution Pipelines and Safety in Tanzania-a case Study

A B Lukonge[1] and X Cao[1*]

[1]College of Pipeline and Civil Engineering, China University of Petroleum (East China), Qingdao, 266580, China

*corresponding author: 15053208836@163.com

Abstract. Energy security is the first priority in any society worldwide. Effective distribution of clean and safe energy is still a challenge for researchers in the energy industry. This study aims to design a natural gas distribution network for supplying clean and safe natural gas for transport, residential, industrial and commercial needs. A case study in Tanzania, for the Tanzania Petroleum Development Corporation (TPDC) natural gas distribution pipeline. Effective pipeline network will reduce the consumption of electricity, coal, and charcoal. For safety purpose, the gas pipeline leak detection system is necessary, which is discussed and proposed in this paper. The old traditional method used for distributing gas energy to the domestic users in Tanzania is through liquefied petroleum gas (LPG) cylinders. This method has numerous drawbacks as cases of accidents are frequently reported.

1. Introduction

Oil and gas in Tanzania have been discovered for more than a half century. In 1974 the first discovery of natural gas was sited at the Songo Songo Island followed by Mnazi Bay in 1982. In 2004 and 2006 the production of natural gas at Songo Songo Island and Mnazi Bay for commercial uses began. Since 2010, numerous discoveries and on-going geological investigation have led to substantial amounts of natural gas both onshore and offshore (see figure 2) [1],[2],[3], [4],[5],[6].

These discoveries have brought more challenges both to the government and the basic company of Tanzania Petroleum Development Corporation (TPDC) about how to manage the fast increasing of projects activities of oil and natural gas industry. The challenges consist of Policy, legal and institutional frameworks to manage oil and natural gas industry successfully [3].

In most parts of the cities, people use the liquefied petroleum gas (LPG) cylinders [7] as a source of gas energy. Although the LPG cylinders in cities can be sold from local shops, there are challenges for consumers about how to fight a fire explosion when it comes to their property and environment in general. This is due to lack of knowledge and protective equipment to consumers. These LPG cylinders have several disadvantages; the basic one is the incorrect method of keeping them at homes which can cause a fire.

The current project is investigating the possibility of constructing a natural gas distribution pipeline network from the city gate centre (Kinyerezi-Dar es salaam) to power plant generation, households, and commercial users.

The project is aiming at using the discharged natural gas from the transmission pipeline from two gas processing plants namely Madimba gas processing plant (Mtwara region) and Songo Songo gas processing plant (Lindi region). Dar es Salaam households, transportation and commercial users will be considered in the designing of the distribution gas network.

This study aims to provide technical designing suggestions for natural gas distribution pipelines in Dar es Salaam city to deliver clean and safe energy for transport, residential, industrial and commercial needs. This proposed system will replace the old system of using LPG cylinders. Also will reduce the use of electricity, charcoal, wood and other fuels . In addition the risk of pollutions and danger to human life and the environment will be minimized.

2. Background of a Master Plan Project.

The national natural gas infrastructure project (NNGIP) was constructed in 2015 from Mtwara region to Dar es Salaam [8]. This natural gas pipeline infrastructure is owned 100% by the Tanzania government under Tanzania Petroleum Development Cooperation (TPDC). It consists 290km, 36" (onshore pipeline) from Mnazi Bay Madimba-Mtwara region to Somanga Fungu-Lindi region, 25km, 24" (offshore pipeline) from Songo Songo to Somanga Fungu-Lindi region, 197km, 36"(onshore pipeline) from Somanga Fungu-Lindi region to Kinyerezi-Dar es salaam region and 30km, 16" (onshore pipeline) from Kinyerezi to Tegeta-Dar es salaam, which make a total of 542km both onshore and offshore natural gas pipeline network up to now as shown in the figure 1. The feasibility study for the natural gas distribution pipelines in Dar es Salaam city was done in 2007,the project aim was to connect industries, institutions, transportation and households [9]

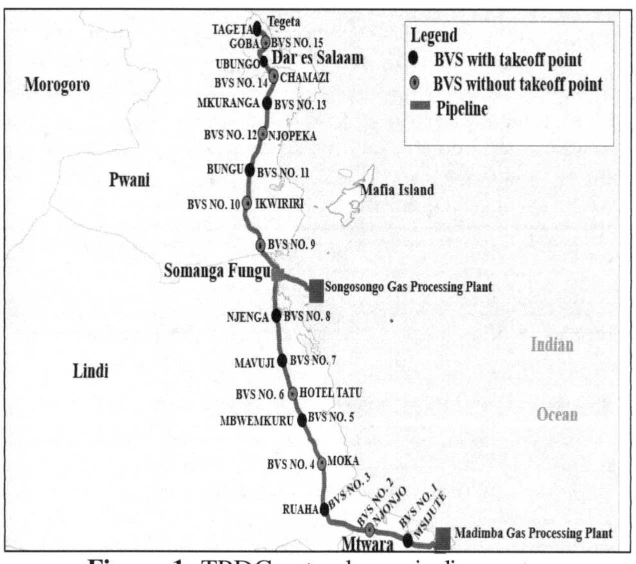

Figure 1. TPDC natural gas pipeline route

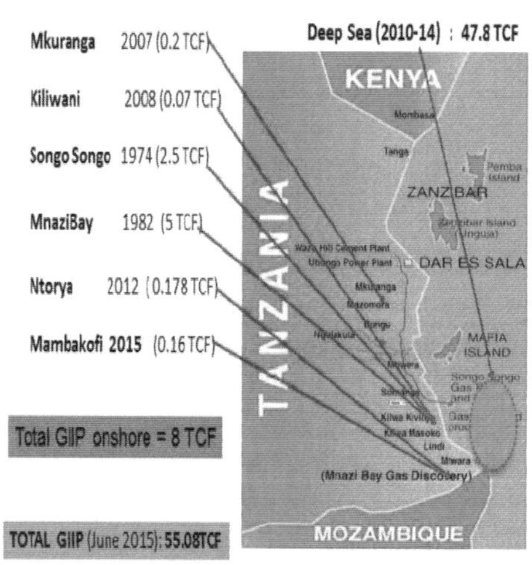

Figure 2. The quantity of discovered natural gas in Tanzania from 1974 to 2015

3. Tanzania energy consumption

Energy is essential in achieving economic prosperity and advances in social and overall human development. Energy has evolved to match with modern human development and requirement.

Biomass is the basic energy supply in Tanzania. This is because about 85-95% of the population of Tanzania does not have access to electricity hence many households use wood and charcoal as their source of energy for cooking and heating [10]. Hence biomass contributes up to 90% of overall primary energy consumption in Tanzania. Other energy sources are oil and gas 8%, electricity 1.5% and coal, wind and solar 0.5% [11] as shown in figure 3 below.

In Tanzania solid biomass in the form of wood and charcoal are the basic source of energy for cooking and heating. The wood used in rural areas is 26 million cubic meters (m^3). Among this, 24 million m^3 is used in cooking for households, 2.03 million m^3 is used in rural areas for small and medium-sized companies and 14.4 million m^3 is used in urban areas - especially as charcoal. According to the United Nations Convention on Climate Change, only 4% of the biomass used is sustainable, outside of the 90% of solid biomass energy is used by consumers. Most biomass requirements are for household used for cooking and heating, and the remainder (10%) is used by home-based companies, businesses, institutions and industries.

Charcoal is one of the main sources of energy for a household in urban areas, with 20% of the total energy consumption. Dar es Salaam city alone uses nearly half of Tanzania's annual charcoal produced which is about around 500,000 tons per year [12]. Therefore, this causes deforestation around 100,000 hectares of trees per year. Thus more efforts are required to decrease deforestation for the sustainability of the environments.

4. Proposed system-Detail design drawing

Switching from traditional energy sources to natural gas usage requires intensive safety care and comprehensive design of gas distribution pipelines. The gas operation can have negative effects on the environment, health and safety of people. Obedience to proper and good industry standards and practices will provide a significant in preventing such negative effects. It is essential to design and plan for remedial measures prior to any operation and management. Regulations of these operations and managements are essential to make sure that activities in the natural gas industry are performed by taking environment, health and safety matters into consideration [2]. Furthermore, the sophisticated gas leakage detection system for distribution pipeline network and equipment monitoring is the best way for preventing or reducing the negative effects to the environment, pipeline network and people for any happening of gas leakage.

The natural gas distribution pipeline schemes should be designed carefully with the required operation standards under the consideration for the economy, safety, and uniformity of pressure issues. The distribution lines must be properly-looped within the particular area and in all outlying areas.

The proposed system should fulfil the existing limitations of the natural gas distribution system in Dar es Salaam city. The proposed additional design and equipment are as explained below in figure 4. Furthermore, we propose real-time gas pipeline leak detection monitoring. Therefore, surface acoustic wave sensor network technology is the best safety measures technology for real-time distribution gas pipelines detection monitoring system.

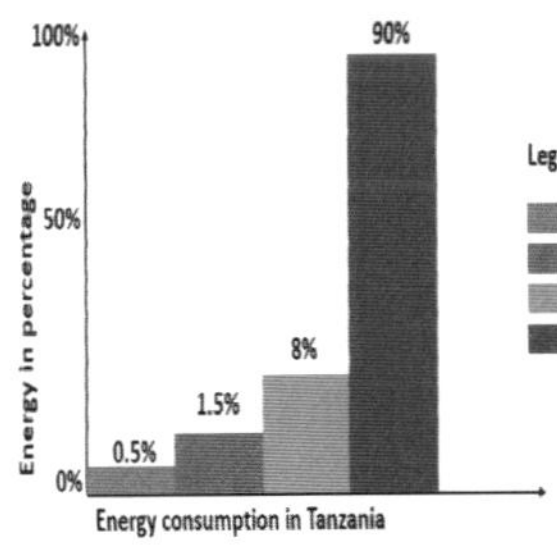

Figure 3. Primary energy supply in Tanzania

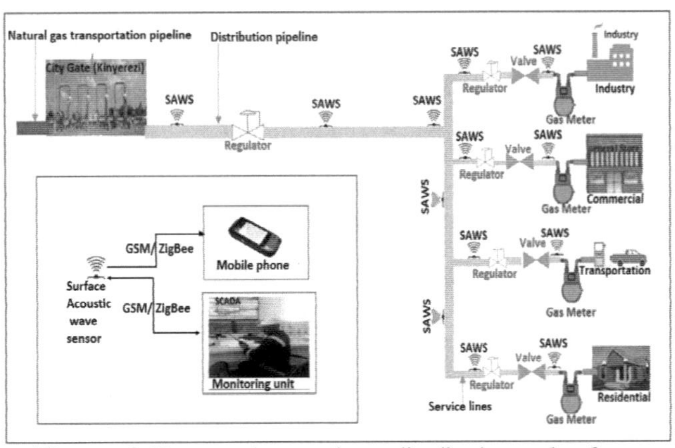

Figure 4. Proposed natural gas distribution and safety system

5. Safety and Monitoring

Each distribution system, gas-meters and other connected devices must be provided with proper pressure regulating devices or acoustic sensors for safety monitoring against any accident that may happen.

When a leakage happens on the pipeline it causes changes to physical parameters on the surface on the pipeline. The surface acoustic wave (SAW) sensors will be used for sensing, identifying and measuring those physical parameters like temperature, pressure, torque, acceleration, tire-road friction, humidity acoustic wave, etc. The sensor will be linked with the remote terminal unit where signal processing will be taken place. The analysed signal from the remote terminal unit (RTU) will be broadcasted to assigned SCADA and mobile phone at a real-time. SCADA and mobile phone will be used as a real-time monitoring system, they will be received all the status of the pipeline distribution system. When a leak happens in the distribution pipeline system the SAW sensor will detect the leak and act as a remote terminal unit by broadcasting the information to SCADA and mobile phone via a global system for mobile communication (GSM) or ZigBee technology. The engineer/operator can receive at real-time that information at the control room and he/she can take quick measures either to shut-down the corresponding valve or the corresponding service line to avoid the serious damages to system, environment and to people. Also through the mobile phone the Engineer can receive the information at real-time anywhere he/she will be and can provide the way forward to operators in the control station.

6. Conclusion

Through the current study we can conclude as follows:

- The supplied natural gas from the distributed pipelines will be safer used by customers, also the environment will be highly protected from the risks of the pipelines leakages.
- Through using cheaper natural gas for cooking and heating people will be economically benefited rather than using LPG
- The use of the electricity will decrease for cooking and heating purposes and hence increase the income to the people.
- The transportation method of using tracks for liquefied petroleum gas (LPG) cylinders will be replaced with this distribution pipeline system; hence natural gas energy can easily be obtained for homes and commercial purposes.
- Also this system can be implemented to other regions and cities where natural gas transmission line passes.
- Through this system, the environment will highly be protected against the risk of deforestation and pollution. Furthermore, users will be also protected as well against any problems that may come after gas leakages.

Amongst the advantages that will be offered by the natural gas distribution system is the capability of supplying large amounts of energy to residential, transportation and commercial users which is more efficiency, reliable and cheaper compared to another source of energy like LPG, Charcoal, kerosene etc.

From this case study carried out at TPDC, we can say that the limited natural gas distribution network and the cost are the key barriers which must be taken into consideration so that to increase the service to the customers and to integrate the market. Finally, if a natural gas distribution scheme will be designed and operated properly, it will drive the development of Tanzania and her citizens abundantly. The clean energy gas for transport, residential, industrial and commercial uses will be easily obtained. This will reduce the demand for electricity, coal, charcoal, fuels, etc.; furthermore the risk from pipeline leakage will be reduced.

7. Recommendations

This chapter identifies some of the works that need to be done in order to provide a basis for the Government to facilitate the implementation of the Natural Gas Utilization Master Plan. This includes:-

- Integration of the master plan project so that any customer can be able to access the natural gas services in their area taking into account that the population of Tanzania is increasing year after year. Also this project should move to another region so as to benefit all Tanzanians and increase the national economy.
- Establishing of the Monitoring system in this system is necessary. Through this, the system itself, the environment and human life will highly be protected against any accident that will be happening in any circumstances. The monitoring system should be monitored wirelessly to enhance real-time monitoring.
- The government, along with other natural gas stakeholders should provide education on the customers and the people in general for basic handling their natural gas appliances and how to protect themselves when gas leakage happens.

8. References

[1] Sospeter M. Muhongo 2013 *The natural gas policy of Tanzania* (Ministry of Energy and Minerals. Dar es Salaam, October)

[2] Muhongo S M 2014 *Local content policy of Tanzania for oil and gas industry* (Ministry of Energy and Minerals. Dar es Salaam , April)

[3] Kinyondo A and Villanger K B H 2016 *Local content requirements in the petroleum sector in Tanzania: A thorny road from inception to implementation?* (REPOA. Dar es Salaam, August)

[4] Boma K J 2013 *The Natural gas sector in Tanzania . Suggestions for a better framework to benefit the Country* (University of Lapland)

[5] Bishoge O K, Zhang L, Mushi W G, Suntu S L and Gregory G 2018 An overview of the natural gas sector in Tanzania - Achievements and challenges *J. Appl. Adv. Res.* **3** 108–18

[6] Inegeja S M 2017 Energy Policy-Tanzania *The Knowledge Co-creation Program, Tokyo Japan.* (Tokyo Japan. 28th June – 29th July)

[7] Amanam U U 2017 Natural Gas in East Africa : Domestic and Regional Use *The Stanford Natural Gas Initiative. Stanford University,May 9 & 10. CANADA* (Stanford University,May 9 & 10. CANADA)

[8] Bofin P, Pedersen R H and Bofin P 2017 *Tanzania's oil and gas contract regime, investments and markets* (DIIS · Danish Institute for International Studies)

[9] Inegeja S M 2017 Status of oil and gas activities in Tanzania *East African Power Industry Convention* (Dar es Salaam. Ministry of Energy and Minerals)

[10] Jonathan Demierre, Morgan Bazilian, Jonathan Carbajal, Shaky Sherpa V M 2014 *Potential for regional use of East Africa's natural gas* (Columbia University)

[11] MSYANI E C M 2013 Current status of the energy sector in Tanzania *Executive exchange on developing an ancillary service market. USEA – WASHINGTON DC* (USEA – WASHINGTON DC, 25th February-2nd March)

[12] Muhongo S M 2017 Renewable Energy. Renewables Readiness Assessment. United Republic of Tanzania *International Renewable Energy Agency, Abu Dhabi* (Abu Dhabi, UAE, 7-10 November)

Acknowledgements

This a case study paper was funded by the National key research and development program of China (No.2016YFC0802302), National Natural science Foundation of China (No.51874340) and TPDC-GASCO, Dar es Salaam-Tanzania.

IOP Conf. Series: Earth and Environmental Science **342** (2019) 012020 doi:10.1088/1755-1315/342/1/012020

Analysis of Lignite Character in Inner Mongolia China

X Y Chen, Y F Zhang* and Q Zhou
Inner Mongolia University of Technology, Hohhot 010051, P.R.China

environzyf@sina.com

Abstract. The character of indigenous lignite was measured by proximate analysis, ultimate analysis, coal ash analysis, coal true relative density, specific area.et al. The results showed that the internal moisture content is 8-9%, volatile matter content is more than 45%, the oxygen content is above 20% while the sulphur content less than 0.5%.The main ash content is SiO_2 and Al_2O_3 , ash fusion point is above 1200 degree centigrade. The calorific value is between 12~19MJ/kg. Macropore and mesopore is the pattern in the lignite.

1. Introduction

The lignite coal regions of Inner Mongolia account for 75% of the total lignite quantity in China [1]. The lignite was formed in the Jurassic era, often be considered as a low rank coal with characteristics between coal and peat, had low calorific value and high moisture content .Its low heating value limits its wide utilization. Now the lignite mainly used in electric power stations near coal-mines[2].

The character of the lignite influence the gravity separation performance, combustion, thermal dissolution and so on[3~5].

The purpose of this work is to investigate the chemical composition, physical property and structural characteristics of the lignite. The character of the lignite is the main internal cause that affects combustion and pyrolysis performance.

2. Experimental

Lignite was supplied from Yiminhe(Y),Pingzhuang(P) and Zhalainuoer(Z) coal mine in Inner Mongolia in China. The sample was dried, ground and sieved to make the particle size is less than 75 μm with a 5EPC1×100 pulverizer . The Proximate analysis of the lignite was performed by a 5E-CHN2000 instrument, and ultimate analysis was carried out with a 5E-MAG6700 element analyzer and 5E-IRS3000 sulfur analyzer. A 5E-KCIV calorimeter was utilized to determine the calorific values. Ash fusion point and ash content were analysis by Carbolite ash fusion point determination meter and EDAX Inc XRF. True relative density was analysised by AccuPyc 1330 fully automatic. True density analyzer and the specific surface area was performed by ASAP2020 Specific surface area and aperture tester. IRTracer-100 FTIR was used to determination the functional groups of lignite. The crystalline structure of lignite was tested by XRD-7000.

3. Results and Discussion

3.1. Chemical composition of the lignite

IOP Conf. Series: Earth and Environmental Science **342** (2019) 012020 doi:10.1088/1755-1315/342/1/012020

Moisture content data are summarized in table 1. Proximate analyses data, ultimate analyses data and calorific value are briefly summarized in table 2. P and Z data have been provided in our previous research[6]. Ash content and ash fusion point was shown in table3 and table 4.

As seen in the table 1, the total moisture content is from 17 to 30 percent, while the internal moisture content is closed to 8-9 percent. High total moisture content makes lignite difficult to be transported[7].

Table 1. Water content of received basis samples.

sample	external water/%	internal water/%	total water/%
Z	21.58	8.15	29.73
P	8.21	9.46	17.67
Y	11.01	9.20	20.21

Table 2. Proximate analysis and ultimate analysis of the samples.

sample	M_{ad}/%	V_d/%	FC_d/%	A_d/%	$Q_{net,ar}$/MJ kg^{-1}
Z	4.90	39.92	48.29	11.78	16.84
P	6.28	32.73	45.39	21.89	18.94
Y	8.88	37.06	47.82	15.12	17.30
sample	C_d/%	H_d/%	O_d/%	N_d/%	S_d/%
Z	63.07	3.88	20.52	0.60	0.14
P	55.67	3.54	17.77	0.66	0.48
Y	60.26	3.40	20.54	0.59	0.09

ad – air dry basis, d – dry basis, M – moisture, VM – volatile matter, FC – fixed carbon, A – ash, Qnet,ar – net calorific value

It can be seen from table 2 that the lignite can be characterized by its low calorific value. High ash content and high moisture make the lignite difficult to be used as power coal, while the sulphur content less than 0.5%. The calorific value of the lignite is in proportion to the fixed carbon content.

Table 3. Ash fusion point of the sample.

Ash sample	Ash fusion point/℃ DT	ST	FT
Z	1092	1202	1302
P	1138	1418	>1500
Y	1132	1198	1248

Table 4. Ash content.

	%								
	Al_2O_3	SiO_2	SO_3	K_2O	CaO	TiO_2	MnO	Fe_2O_3	MgO
Z	10.65	52.39	9.39	0.30	21.38	0.73	0.61	4.13	0.42
P	17.11	72.04	1.38	2.57	3.11	1.19	0.04	2.37	0.19
Y	6.18	62.73	4.92	0.15	18.88	0.40	0.10	2.45	4.19

Table 3 and table 4 showed that SiO_2 is the major ingredient of the ash. The previous research indicated that the ash fusion point values decrease with the increase of SiO_2 Mass percentage when the SiO_2 Mass percentage value between 45% to 60%, SiO_2 plays a role of flux[8,9]. There was no certain regulations of the SiO_2 Mass percentage value and ash fusion point values when the SiO_2 mass percentage above 60%, while the ash fusion point value was high when the SiO_2 Mass percentage value exceed 70%, that is consistent with the experimental results.

Al_2O_3 content has an effect on ash fusion point value. Al_2O_3 has a solid crystal structure and the fusion point is 2025 ℃.The ash fusion point value goes up regularly with the Al_2O_3 content increased when its content higher 15%.

3.2. Physical characteristics of lignite

True relative density of the lignite was summered in table 5, The specific surface area and aperture were shown in table 6. The adsorption isotherm of lignite was shown in figure 1.

Table 5. True relative density of the lignite.

sample	True relative density (g/cm³)
Z	1.4082
P	1.4706
Y	1.5280

It is evident from Table 5, the true relative density value is higher than the existing test results, the reason is mechanical crushing make the pore be broke. The true relative density value is related to the porosity, the high true relative density value corresponding to the low porosity.

Table 6. Specific surface area and BJH mean pore size.

sample	Z	P	Y
BET/$m^2 \cdot g^{-1}$	2.78	5.97	3.05
BJH/nm	40.84	18.21	52.73

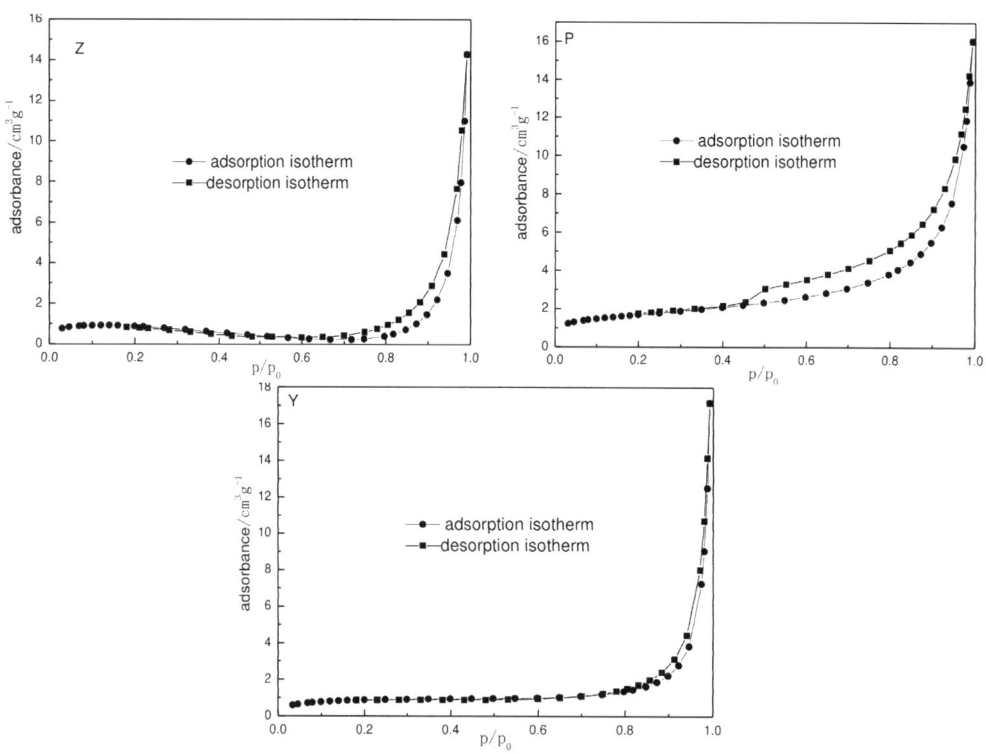

Figure 1. Adsorption-desorption isotherms of the sample.

Dr. *Zhao*[10]mentioned adsorption isotherms were divided into five types according to the pore distribution, generally the isotherm of coal type belongs to II type. It can be seen from Figure 1 that the linetype belonged to III type. That means the major hole pattern in the lignite is mesopore and macropore, the results are agree with table 6.

3.3. Structural characteristics of lignite

FTIR characterizes the surface functional groups of coal. The thermal decomposition products is relationship with the functional groups of coal, the coal pyrolysis is the functional-groups decomposition. X-ray diffraction is an important method to determine crystal structure at present. Organic matter in coal is a substance between crystal and amorphous. The result can represent the arrangement of carbon atoms in coal.

Infrared spectra of lignite was showed in Figure 2. Transmission occurs at the frequency of 3695 cm^{-1} and 3620 cm^{-1} belongs to hydroxy free stretching vibration. There is a wide transmission peak from 3500 cm^{-1} to 3100 cm^{-1} , It can be assigned to hydroxyl especially phenolic hydroxyl group. There is a number of hydrogen bond reciprocity because of the associated with the phenolic hydroxyl group, which make the high oxygen-containing groups number. The weak acromion peak at 2900 cm-1 and 2870 cm-1 is attribute to the stretching vibration of -CH in the alkyl chain, which indicating that aliphatic structure exists in lignite. Peak about 1610 cm^{-1} is the contribution of the C=C bond on the aromatic ring. It is concluded that there are aldehyde and ketone groups in the lignite.

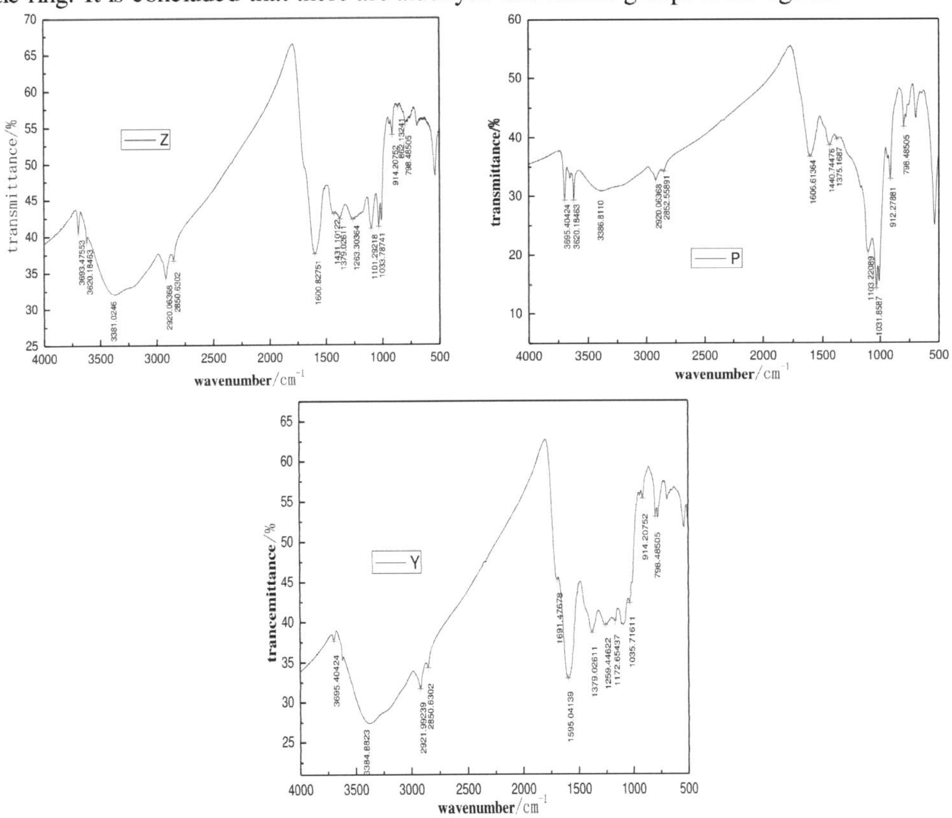

Figure 2. FTIR of the sample

The XRD curve was shown in Figure 3.

It can be seen from figure 3, the peak of the curve is wider and irregular, the carbon atoms are randomly arranged in organic molecule of the lignite with low degree of the crystalline and low condensation of aromatic rings.

d002 represents the distance between the aromatic layers, the high value the big of the layer. That means the loose construction of the lignite, the value of d002 was shown in table 7.

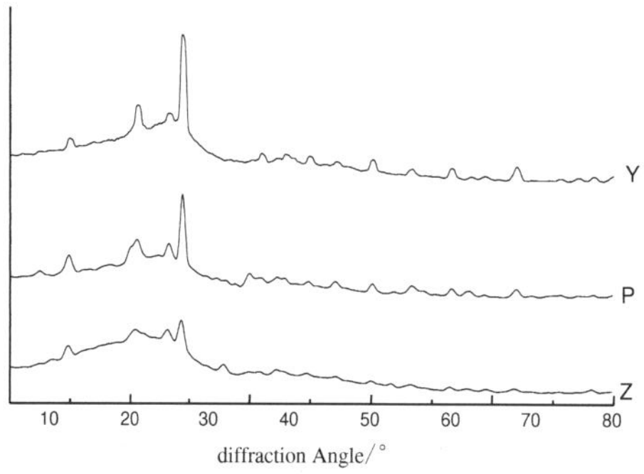

Figure3. XRD curves of the samples.

Table 7. Structural parameters of microcrystals of the sample.

sample	diffraction angle/°	d_{002}/nm
Z	26.59	0.33492
P	26.59	0.33491
Y	26.71	0.33341

4. Conclusion

The chemical composition and physical property s of lignite were studied to provide theory gist for study combustion and pyrolysis behaviour. Our results showed the following conclusions.

The moisture content is above 30% of the lignite, with high volatile that the value is above 45%. The lignite has the high oxygen content and the low sulphur content. SiO_2 is the main component of the ash, the other is Al_2O_3, which make the ash fusion point value above 1200℃

Particle size of the lignite differ widely, the major hole pattern in the lignite is mesopore and macropore, that make the lignite has the loose construction.

There are more oxygen-containing functional groups in the lignite .Lignite has the low degree of aromatization and carbon atoms in organic molecules are disorder.

The directions of our future study will focus on emission rules of combustion pollutants related with the character of the lignite.

5. References

[1] Yin L Q 2004 *China. Coal Science and Technology*. **32** 12-14
[2] Jiang F Y, Wei M Y, Chen D L, Ye T and Wen C 2018 *China. Power System Engineering*. **34** 19-22+25
[3] Zhu X N,Wang Q L and X T 2019 *Energy Sources Part A*. **41** 1527-33
[4] Haykiri-Acma H, Cekic Y, Yaman, S. 2019 *Energy Sources Part A*. **41** 1326-35
[5] Wu F P, Yan J, Zhao Y P 2019 *Energy Sources Part A*. **41** 1132-39
[6] Chen X Y, Zhang Y F, Zhang Q C, Li C P and Zhou Q 2015 *Thermal Sci*. **19** 801-11
[7] Yu J L, Tahmasebi A and Han Y N 2013 *Fuel Process Technol*. **106** 9-20
[8] Chen M Q, Yu D and Wei Y H 2016 *Power Technol*. **286** 39-47
[9] Li F H and Fang YT 2015 *Energy and Fuels*. **29** 2979-86
[10] Zhao Z G 2005 *Application principle of adsorption*(Peking: Chemical Industry Press) p71- 73

Acknowledgments

This work was financially supported by National Natural Science Foundation of China (21066009, 21467020), Natural Science Foundation of Inner Mongolia (2017MS(LH)0206).

IOP Conf. Series: Earth and Environmental Science **342** (2019) 012021 doi:10.1088/1755-1315/342/1/012021

Fine Dust Removing by Using Li+ Batteries and High - performance Supercapacitor Parallel Connection

Jin-Soo Bak and Hee-Je Kim*

Department of Electrical and Computer Engineering, Pusan National University, South Korea

*Corresponding Author E-mail: heeje@pusan.ac.kr

Abstract. One of the most serious problem in the world, Fine dust issue is going to happen in whole world. 'Fine dust' is defined under these standard, if its diameter particle is less than 10um, it is called as 'fine dust', 2um, it is called as 'super fine dust'. Actually, 'Fine dust' is assigned as 1st level hazardous substances by WHO (World Health Organization). In this paper, I suggested new way to solve this kinds of problem by using Li+ batteries and high-performance supercapacitor. A lot of kinds energy source such as fossil fuel, one of the reason of environment problem can be changed to eco-friendly energy such as solar power. We can solve two factors o(cause / result) of fine dust by using eco-friendly energy source.

1. Introduction

Nowadays, the amount of demand of fossil fuels for personal vehicle and factory has been increased for a long time. This kinds of wave has brought serious problem in recent. The components of fine dust (Figure 1) are composed of Sulfide, Nitrate and Carbon, Metal things. Usually, we used to think 'Fine Dust' as just one kind of particulates like sand storm but it is needed to call as micro metal dust. In the case of dust and sand storm, it's a natural weathering phenomenon. However, in the case of the fine dust, too much soot and smoke happened in the air so that it is getting clutter with sulfide, water and etc. So it has totally different mechanism with just dust and 'Fine Dust' as I defined [1].

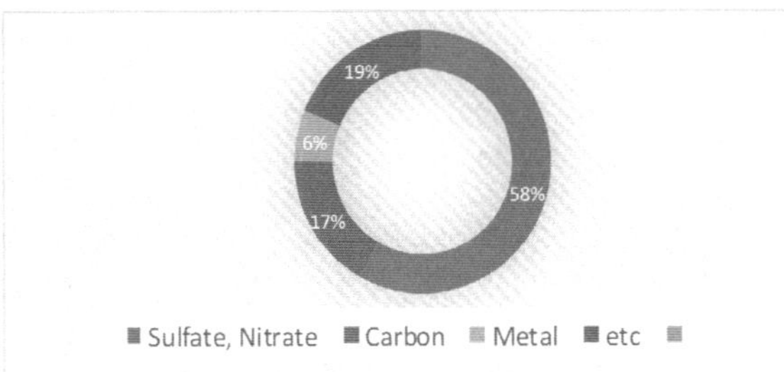

Figure 1. Components of Fine Dust

In other words, super-micro metal powder is always around you and even you don't know when you have some powder in air every single breath. The particle size of fine dust is a very small as less than few micro meter so it's difficult to recognize and remove.

There are 3 ways to remove the fine dust, it is mainly using a filter, centrifugal force and electrostatic dust collection. One of this, the dust collection way is not efficient for the small particles. Further more, the way of using a filter is non-economic because it's essential to be placed somewhere which has enough space. In this study, I suggested the new design of fine dust removing by using Li+ batteries and supercapacitor parallel connection. I hope it can be the one of new prototype of technology for fine dust.

2. Experimental

2.1. The principle of fine dust removal

The principle of a corona discharge is based on the electrical discharge by ionizing of a fluid such as air surrounding a conductor that is electrically charged. Naturally corona discharges occur with high-voltage systems unless electric field strength is up to limit. A corona will occur when the strength (potential gradient) of the electric field around a conductor is high enough to form a conductive region, if it's not high enough to cause electrical breakdown or arcing to objects nearby. As it showed in Figure 2, it's often seen as a bluish (or other color) glow in the air adjacent to pointed metal conductors carrying high voltages and emits light by the same property as a gas discharge lamp.

Figure 2. Corona Discharger

The electrostatic precipitator is a device for collecting dust that removes fine dust by attracting various dust particles contained in the air to a dust collecting pole just like Figure 3, A corona discharge occurs when a high voltage is between the (+) pole and (-) pole. When it occurs, ions with charges are generated. (-) Charged ions combine with the fine dust particles in the air (-) making the dust migrate to the (+) pole attached to the electrostatic precipitator. This principle makes it easy to remove various kinds of fine dust particles[2-4].

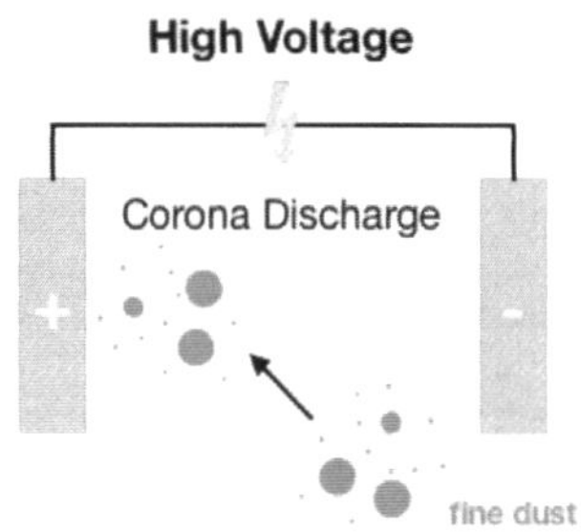

Figure 3. The method to get rid of fine dust

2.2. Design of the circuit and system

Solar energy is one of the green power source and it is no limit. So I designed this system to easy to operate with solar energy. Solar Panel (30W output, 20V), Solar Controller (Output: 12V), Boost Converter (Input : 12V, Output : 40V / Mc34063A IC), Supercapacitor (12V), Li+ Battery (12V) and

Fine Dust Removal System (Input 30-70V) :: **Figure 4.** Total 30-70 voltage in needed to operate this system, the power comes mainly from Li+ batteries charged by solar energy and supercapacitor. However, the solar panel's maximum output voltage is only 20V so it's required to use the boost converter to make output voltage high up to 40V.[5]

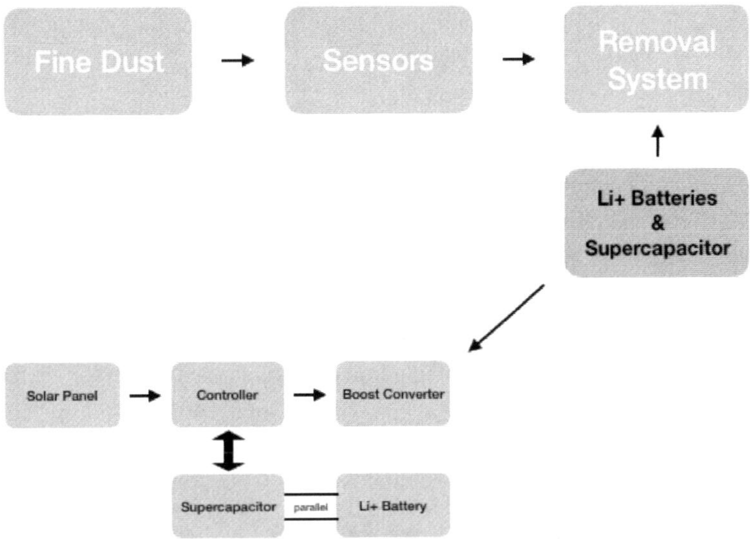

Figure 4. Design of Total Circuit System

2.3. Electrodes

I prepared two different types of aluminum electrodes, one is cylinder shape and the other one is quadrangular prism type. The length is 25cm and the gap distance of separated electrode is 2cm on each electrodes. The reason why I made it two types is it's kind of the first trial so it's needed to compare by changing the structure of system. And I used 3 types of fine dust such as HCHO, TVOC and SMOKE. I compared how much fine dust is removed by this system as I change the structure of system.[6]

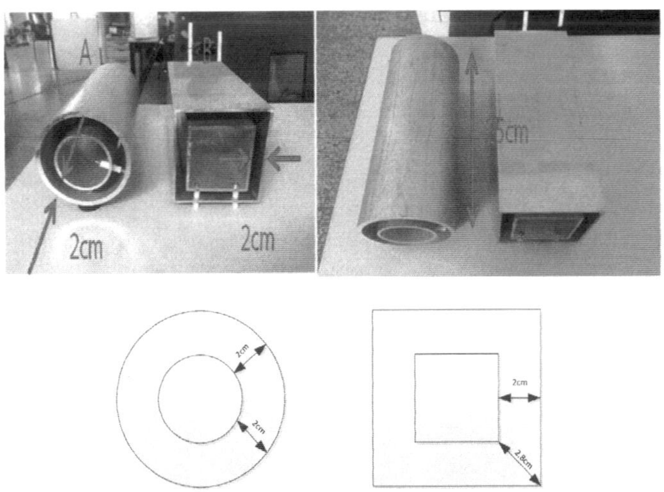

Figure 5. Designed Electrodes

3. Result and Discussion

I did this experimental under this options of measurement as shown in Table 1.

Table 1. Options of Measurement

HCHO/TVOC		PM1/PM2.5/PM10	
Metrics	HCHO&TVOC	**Particle diameter test**	1μm/2.5μm/10μm
Measuring range	0~1.999mg/$\mathbf{m^3}$(HCHO)	**Detection mode**	Density (per liter)
	0~9.999mg/$\mathbf{m^3}$ (TVOC)	**Detection range**	0~999 μg/$\mathbf{m^3}$
Sample Mode	Diffusion type		
Density unit	mg/$\mathbf{m^3}$		

In this study, Smoke, Fine dust, HCHO, and TVOC are measured using an air purifier. This study compares the efficiency of the different type of pipe and the optimal removal time of smoke, fine dust, HCHO, and TVOC. Firstly, the smoke removal experiment is carried out. Smoke contains large dust particles. Therefore, the optimal frequency is determined through the smoke elimination experiment, and the optimal conditions of the other experiment are also examined.

As shown in Figure 6 (a), when the switching frequency is 13 kHz. The circular pipe (A), and the square pipe (B) models are the fastest in removing the smoke. The HCHO, TVOC, and PM 2.5 (particle material) experiments are performed with a frequency of 13 kHz, 40cmx20cmx20cm acrylic box, and input voltage of 30V. Formaldehyde (HCHO) is a flammable colorless gas that has a long-lasting, irritating odor with adverse effects on the human body, such as headache / amnesia and emotional disturbances. If the HCHO density value is > 0.30 mg/$\mathbf{m^3}$, the present air is not good. The data are derived by comparing the A and B pipes, and no air purifier only fan(C model) shown in Figure 6(b). The A model showed the fastest purging. The air is purified within 10 minutes by the A and B models, whereas 100 minutes is needed using the C model (Without using an air cleaning system, only fan). TVOC is a precursor of a photochemical reaction and means a substance harmful to the human body / environment by generating secondary pollutants, such as ozone and aldehydes. The experimental results show that the A model takes approximately 4 minutes and the B model takes more than 10 minutes, whereas, the C model takes 100 minutes shown in Figure 6(c).

If the PM 2.5 concentration is greater than 115 $\mathbf{\mu g/m^3}$, the present air is not clean. The experimental results show that the A and B models cleaned the fine dust at the same time. On the other hand, C model clean the fine dust after 100 minutes shown in Figure 6(d). (Table 2) lists the results of the purification. The experimental results of the A model shows excellent results under all conditions. The A model has a uniform surface different from the B model, and it purifies the air the fastest. In the case of the B model, partial discharge occurs in the pointed portion; hence, it is relatively inefficient. In case of the C model it does not have a dust collector, but it has fan function so whenever it operates the air circulates occur inside. As the time passes due to internal circulation the C model parameter gradually decreases.

Follow the experiment, we can find the A model is most effective to remove the fine dust. In case of the A model has uniform surface between (+) pole and (-) pole. But the B model has not uniform surface. Because of the B model has irregular sections like edges of square. So the B model can`t remove fine dust efficiently. Therefore, we can conclude the A model is most effective.

Table 2. Purification time on A and B

Type of matter	A Model	B Model
Smoke	4(Second)	5(Second)
HCHO	4(Minutes)	10(Minutes)
TVOC	4(Minutes)	10(Minutes)
PM2.5(Fine dust)	4(Minutes)	4(Minutes)

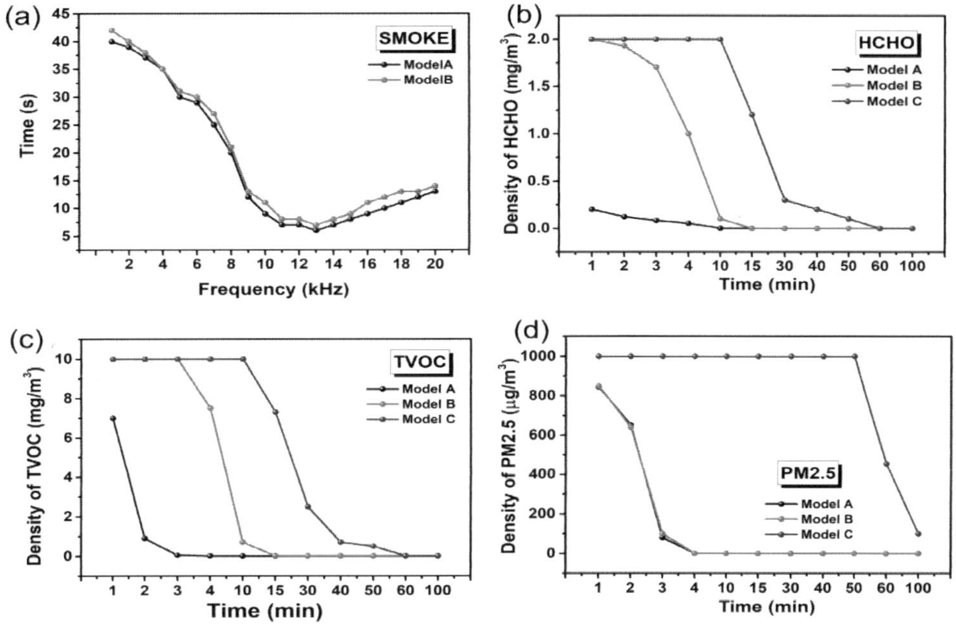

Figure 6. (a)Smoke (cigarette) removing time versus switching frequency; (b) Three different types of HCHO purification time; (c)Three different types of TVOC purification time; (d) Three different types of fine dust purification time

4. Conclusion

In this research, I made a corona discharge system by using a half bridge inverter, step-up transformer and Cockcroft Walton circuit. And I presented here two new types (cylinder, quadrangular prism) for the structure of fine dust removing system. Mainly HCHO, TVOC and SMOKE is measured with two types of electrodes and as a result, the model A was able to remove the fine dust on 30V of input voltage for 4 sec. The HCHO, TVOC and fine dust, all particulate matter was visible but it's going to be invisible so this type of system by using Li+ batteries and supercapacitor is possible to operate outside properly.

5. References

[1] Cristina, S.; M. Feliziani, Industry Applications Society Annual Meeting, Conference Record of the 1991 IEEE 1991, 616-621
[2] Animals see power lines as glowing, flashing bands, research reveals, 2014
[3] Vishay Offers C-stability in X2 Capacitors, CapacitorIndustry.com., 2017-11-22.
[4] L. Xia, L. Huang, X. Shu, R. Zhang, W. Dong, and H. Hou, "Removal of ammonia from gas streams with dielectric barrier discharge plasmas," J. Hazardous Mater., 2008, 152, 1, 113-119
[5] W. F. L, M. Hoeben, F. J. C. M. Beckers, A. J. M. Pemen, E. J. M. van Heesch, and W. L. Kling, "Oxidative degradation of toluene and limonene in air by pulsed corona technology," J. Phys. D, Appl. Pys, 2012, 45, 5,055202.
[6] Khan, M.A.; Ko. B.; Alois Nyari, E.; Park, S.E.; Kim, H.-J, Performance Evaluation of Photovoltaic Solar System with Different Cooling Methods and a Bi-Reflector PV System (BRPVS): An Experimental Study and Comparative Analysis. Energies 2017, 10, 826.

IOP Conf. Series: Earth and Environmental Science **342** (2019) 012022 doi:10.1088/1755-1315/342/1/012022

Enhancing the Performance of Dry Air-Cooled Systems with Wet Geothermal Cooling in Saudi Arabia

Mahmoud Bady[1,2], and Nabil A.S. Elminshawy[3]

[1] Mechanical Engineering Department, Islamic University of Medina, Saudi Arabia

[2] Mechanical Engineering Department, Assiut University, Assiut 271516, Egypt

[3] Mechanical Power Engineering Department, Port-Said University, Egypt

Email: mfbady@yahoo.com

Abstract. Reducing the temperature of the working fluid in the electric generation systems (gas turbine air inlet as well as steam from the steam turbines) requires an effective cooling medium. The present work is a part of a research project conducted in the Islamic University of Madinah, Saudi Arabia, to present reasonable solutions to pre-cool the inflow air to the air-cooled condensers in electric generation systems in KSA, when the ambient air temperature increases above the design air inlet temperature. An extensive amount of measurements has been carried out to investigate the effects of soil parameters on a buried earth-air heat exchanger. Among many parameters, the current research focused on investigating the effects of soil moisture contents and degree of compaction -represented by relative density- on the thermal performance of the buried heat exchanger. Results of the study show that; an average daily basis, a significant induced air temperature drop of 23.5 °C is achieved at the soil moisture content of 30 % and soil compaction of 91.17 %. Furthermore, the effectiveness of considered BEAHE is increased by about 33.34 % when the soil compaction is increased from 26.47 % to 91.17 %.

1. Introduction

The majority of power plants built during the last decade around the world are based on the combined-cycle process for power generation. The new combined-cycle power plants are seen as offering a greater degree of flexibility in the type of cooling system that can be used. Several recently built combined-cycle plants use air-cooled condensers (dry cooling) instead of the traditional condenser/wet cooling system. Still, other plants have hybrid cooling systems.

Kingdom of Saudi Arabia has a long dry summer extending over about seven months, from March to November. The design conditions and the monthly average peak of dry bulb and wet bulb temperatures indicate the hot-dry nature of KSA weather occurs during the hot summer months (April to September) when peak daily temperatures range from 37 to 55 °C. For most of the summer months, the difference between the average DBT and WBT is over 15 oC the relative humidity was between 14.4 % and 20 %. Thus, water cooling condensers (WCC) are expected to perform more efficiently as the sink temperature is lower. In spite of the apparent advantage of the lower power and energy demand of the WCC systems, most of the power plant cooling systems in Saudi Arabia are of the air-cooled types (ACC) type. Due to the shortage of natural and soft water, and rarely rainfall, most of the soft water requirement is satisfied through seawater desalination, which is highly energy intensive. The high cost of water is, therefore, an additional parameter affecting the choice between ACC and WCC systems.

Dry cooling of power plants in Saudi Arabia may be an attractive alternative to wet cooling, particularly where water conservation and environmental protection pose critical sitting issues. However, dry cooling technology may be unable to maintain design plant output during the hottest periods of the year, which are often periods of peak system demand.

Many studies have been carried out to investigate the effects of cooling systems on the cycle performance and energy generation cost. Maulbetsch and DiFilippo [1] stated that the total thermal power plant cost is 5% to 15% higher if dry-cooling instead of wet-cooling is utilized. They also noticed a reduction in the potential annual income may be 1% to 2% that for dry-cooled systems, due to their performance penalties in a hot climatic region. Weifeng et al. [2] stated that the efficiency of the thermal systems utilizing air-cooled condenser is adversely affected when ambient air temperature increases above a certain acceptable nominated temperature. So, the larger the temperature difference between the heat source of the thermal system and the external ambient atmospheric air were simply transferring its exhausted heat to the ambient air directly, the more efficient is the thermal system cooling. Thus, the desirability of lowering the air ambient temperature in the external environment is significant. This consideration gives rise to desirably hybridizing the active air-based units of thermal system alongside with passive techniques (i.e. earth-air heat exchangers) [3].

Outdoor ambient atmospheric air is induced into the buried earth-air heat exchanger (BEAHE) before being introduced to the thermal systems for air cooling purpose, rather than being supplied into the thermal systems directly. When air passes through the BEAHE in the summer season, heat is transferred from the air to the soil, the air is then pre-cooled. Due to the delayed thermal response of soil, the temperature of the soil at a few meters under the earth surface is lower than ambient in summer and remains nearly constant at the average annual level [4-8]. The BEAHE utilizes this temperature gradient between atmospheric air and ground for passive cooling by dumping heat into the soil as a natural sink, as a result, the air is pre-cooled by the soil in the summer season. The outlet pre-cooled air from the buried earth-air heat exchanger can be directly employed for enhancing the thermal system cooling. Buried earth-air heat exchangers are eco-friendly and consume almost no energy. Thus power consumption of the vast number of thermal systems can be reduced substantially by hybridizing it with earth air heat exchanger system especially in hot and humid weather conditions where the summer ambient temperature sometimes exceeds 40 ◦C.

Several configurations for shallow buried earth-air heat exchangers have been proposed as a heat source or sink, two of the most popular configurations are horizontal and spiral coil heat exchangers. Horizontal ground heat exchangers [9, 10] and spiral coil heat exchangers either used directly in the ground or in pile heat exchangers [11, 12]. However, these types of earth-air heat exchangers are more sensitive to atmospheric conditions, and to dynamical behavior of soil moisture contents because they are closely buried to the earth surface. Soil-specific conditions at the site around the BEAHE found to highly impact horizontal earth-air heat exchangers systems designed and performance [13, 14]. In the most recent studies, thermal performance of the earth-air heat exchanger models is always based on heat transfer, without consideration of the soil moisture contents [15, 16]. Only few research works have been accomplished in soil moisture migration caused by buried heat exchangers operation, such as the model designed by Chalhoub et al. [17] to check the effect of soil moisture around the coil was developed. According to a computer simulation by Leong et al [18], the thermal performance of a ground heat exchanger system shows a very sharp decline for dry soil—it is lower by up to 35% with respect to saturated conditions. Alteration of soil moisture content from 12.5% of saturation to complete dryness strongly decreases the ground heat exchanger effectiveness. The soil is usually unsaturated around the buried heat exchanger, in which heat transfer interacts with water transfer in a very complex process. In another investigation by Piechowski [19], the initial value of soil moisture content was significant for accurate simulation or construction of buried heat exchanger. In their experimental investigation that was testing four soil types, an increase in moisture content at a given bulk density increased the thermal conductivity. Furthermore, parameters that affect the thermal conductivity of soil can be controlled externally. Soil density and moisture content are some of these controlled parameters [20].

The main objective of the present research is to investigate experimentally a newly air cooling technique employing geothermal sustainable cooling that will integrate with the traditional cooling or condensation system that already used with the vast numbers of the thermal system, especially in the summer season. In this study, the ambient air is firstly cooled via a buried heat exchanger which assisted by various soil moisture contents to enhance air cooling by damping the heat of the atmosphere through the soil. The produced pre-cooled atmospheric air will then pass to the traditional thermal cooling systems but with low temperature than the hot air temperatures. The thermal performance of such geothermal wet cooling as a new cooling system will be calculated using two parameters; the soil cooling load and the whole system effectiveness.

2. Experimental setup and measurements

The present study is based on the extraction of heat from the hot ambient air in the summer season and then passes it to the soil beneath the earth surface using the BEAHE.

2.1. Experimental test rig and procedures

To evaluate the performance of BEAHE in various climatic conditions and in order to investigate the effects of different operational parameters, an experimental setup was designed and constructed, as shown in figure 1. The test setup consists of four major modules:

- A primary air simulator including a centrifugal air blower of 0.25 kW equipped with a frequency inverter to adjust air flow rate and electrical heating chamber with thermostat was employed. The electrical heating chamber was used to heat the atmospheric ambient air to the required temperature that represents the actual ambient air in the summer season in Madinah. So, the ambient air with various temperatures and flow rates was supplied to the tubes of the buried heat exchanger. A straighter was provided to achieve airflow uniformity and temperature homogeneity.
- A cylindrical truncated drum of 0.5 m diameter with a length of 1.5 m was devoted to containing the selected soil to be tested where the heat exchanger tube was buried. The drum is made of 1.5 mm-thick galvanized steel sheet and aligned horizontally. So, three boreholes were drilled at various depths in the site of Madinah land, at 1, 2 and 3 m from the earth surface.
- The exact soil temperatures were measured in the three various selected depths using cable thermometer. Then, the soil temperature that representing each depth is tested in the lab by filling the truncated drum with the soil collected from the same site around the boreholes and keeps it in a control temperature lab for 24 hrs before each test. This sequence ensures constantly desired soil temperature homogeneity at the begging of each test that represents each depth. The buried tube was constructed of copper with a length of 1.5 m and a radius of 0.02 m and was aligned at the center of the cylindrical soil drum.
- The experimental investigations were conducted in the hottest month (i.e. July) to evaluate the thermal performance of the considered earth-air heat exchanger for cooling the hot ambient air.

By letting hot ambient air flow through the buried tube, the air temperatures in the earth–air–tube system were measured by multichannel K-type thermocouples with a digital display connected at four various radial positions from the center of the tube. These four equally spaced radial positions are referred to as 0.045, 0.090, 0.135 and 0.180 m from the center of the buried tube. A Kanomax multi-function hot-wire anemometer Model 6501 was used to measure the airflow and temperature at the inlet and outlet of the buried tube. By measuring the ambient air temperatures at the inlet and outlet of the heat exchanger tube at 30 min intervals, hourly cooling potential by BEAHE can be calculated for the various studied working parameters according to the equation:

$$\dot{Q} = \dot{m} c_{\mathrm{p}} (T_{\mathrm{inlet}} - T_{\mathrm{exit}}) \tag{1}$$

The effectiveness of the BEAHE system is given as:

$$\varepsilon = \frac{T_{\mathrm{a)in}} - T_{\mathrm{a)out}}}{T_{\mathrm{a)in}} - T_{\mathrm{w}}} \tag{2}$$

where T_w is the tube wall temperature at the start of each experiment, $\dot{Q}$ is the cooling potential (W.h), $\dot{m}$ is the air mass flow rate through the tube assembly (kg/s); c_p is the specific heat capacity of air (J/kg. K); $T_{a)in}$ is the air temperature at the inlet of BEAHE (^{o}C); $T_{a)out}$ is the air temperature at the outlet of BEAHE (^{o}C).

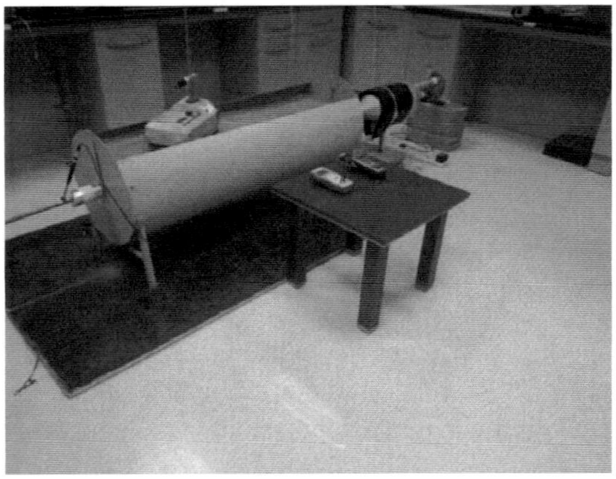

Figure 1. Photograph of the test rig set-up

The current research focused on the effects of soil moisture contents and degree of compaction represented by relative density (DR %) on the thermal performance of buried tube heat exchanger. For the interpretation of the test data, both moisture contents and the degree of compaction of the unsaturated tested soil were controlled. Three moisture contents were prepared; 9.87, 21.3, and 30 %, where each level being tested at three degrees of soil compaction of 26.47, 70.58, and 91.17 %. It should be noted that; it was difficult to prepare indoor soil with a higher degree of compactions, especially at the high water contents. The soil was firstly placed in an oven and allowed to dry for 24 hours and is then left in a dry place before being used, through this process the mass and density of dry soil can be evaluated. According to the required soil moisture content, a fixed mass of the dried soil was thoroughly mixed with the desired amount of water to produce soil with different initial moisture of content.

2.2. Tested Soil Properties
It is expected that the properties of soil affect the performance of the considered earth-to-air tube heat exchanger as natural heat dissipation sink. So, undisturbed natural soil samples collected from the site of Madinah land around the Islamic University campus was subjected to laboratory measurements to determine the bulk density, porosity, and particle size distribution. It is dusty brown poorly graded sand its physical properties according to the current research scope are as follows.

2.3. Compaction Characteristics
This multidimensional study fuscous on utilizing soil heat dissipation properties which greatly depends on the voids present in the soil. The number of voids present in the soil can be represented by the degree of compaction (relative density). The relationship for determining relative density from void ratio is given by:

$$DR(\%) = \frac{e_{\max} - e}{e_{\max} - e_{\min}} \times 100 \tag{3}$$

where e_{max} is the maximum void ratio, e_{min} is the minimum void ratio, and e is the void ratio at given compaction.

The void ratio is calculated on the basis of simple mass-volume relationships. Whereas, maximum and minimum void ratios are determined in laboratory respectively as per ASTM D 4253 and ASTM D 4254. The maximum and minimum void ratio values for Madinah Sand are given in Table 1.

Table 1. The maximum and minimum void ratios for Madinah Sand

Parameter	Value
e_{max}	0.97
e_{min}	0.63

2.4. Compaction Levels for Model Test:

To obtain different bulk densities, the soil was compacted in the truncated drum by a soil compaction hammer. The hammer of an approximate weight 2 kg, was dropped into the soil layer surface inside the drum from a height of approximately 0.30 m. To achieve different levels of bulk density, one used zero drops for the loose soil, 10 drops for the medium soil, and 25 drops for the compacted dense soil. Three compaction levels have been selected for the analysis of heat dissipation properties of the investigated BEAHE. These levels have been abbreviated as T_1, T_2, and T_3. The properties of these levels are summarized in Table 2.

Surface moisture-density gauge Model 3430, manufactured by TROXLER was employed to quickly and precisely measure the compaction level moisture and density of soil and soil bases without using destructive methods. Using direct transmission or backscattered gamma radiation, this gauge measures the density of materials by counting the number of photons emitted by the cesium-137 source.

Table 2. Properties of the three compaction levels for Madinah tested soil

Sand Nomenclature	Compaction Level	Specific Gravity	Void Ratio (e)	DR%
T_1	Loose	2.67	0.88	26.47
T_2	Medium	2.67	0.73	70.58
T_3	Dense	2.67	0.66	91.17

The daily average improvements in the cooling loads (% Q_c) across the BEAHE for wet soil with different moisture levels and compaction levels compared with dry soil is an important performance parameter. It can be estimated through the formula:

$$\%Q_{\text{c−mprovment}} = \frac{Q_{\text{c)wet}} - Q_{\text{c)dry}}}{Q_{\text{c)dry}}} \times 100 \qquad (4)$$

2.5. Ground temperature measurements

The mean soil and the ambient air temperatures were measured are shown in figure 2 for the months of June to September in the site around the Islamic University campus at three depths inside the soil: 1, 2 and 3 m. Clearly, the soil temperature decreases significantly beneath the ground surface compares to the ambient dry bulb temperature. It is also seen that, while the mean air ambient temperature fluctuated around 44-49 ℃, in the summer months during the day time, the temperature at 1-3 m depth remained virtually constant at 27-29 ℃. A maximum difference of about 18 ℃ was observed between the soil temperature and the dry bulb temperature of the ambient air.

2.6. Uncertainty in the measurements

In the current investigation, the minimum measured values of induced air velocity, air temperature, soil relative density, and soil moisture content are 10 m/s, 20 °C, 26.47 % and 9.87 % respectively. Least count of measuring instruments for induced air velocity, temperature, soil relative density, and soil moisture content are 0.015 m/s, 0.1°C, 10% and 5%, respectively. Therefore, uncertainties in the measurement of induced air velocity, temperature, soil relative density, and soil moisture content are estimated as ± 0.60%, ±0.5%, ±0.37%, and ±0.50% respectively.

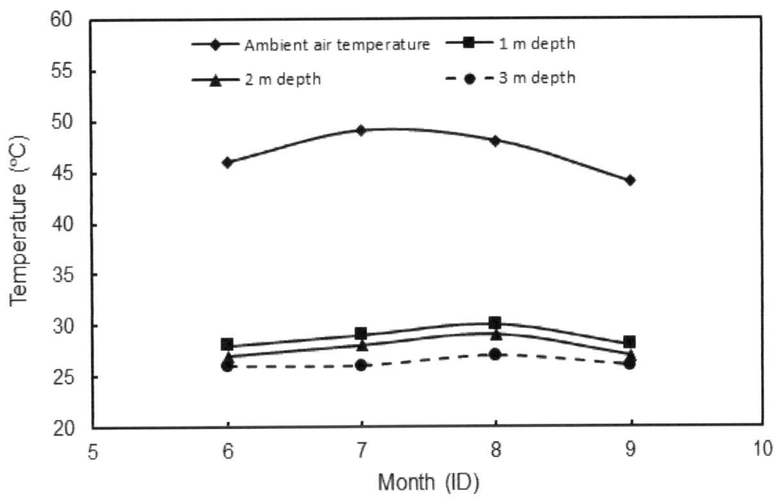

Figure 2. The ambient air temperature and soil temperature at different depths.

3. Results and discussions

Among many factors, the soil relative density and the soil moisture content are selected to study their effects on the proposed heat exchanger performance.

3.1. Effect of soil moisture contents and compaction level on the cooling load:

The effect of soil moisture contents on the temperature drop across the considered BEAHE for various soil relative densities is illustrated in figure 3 through figure 5. Tests were performed for wet soil with three different levels of soil moisture contents of 9.87, 21.3, 30% and compared with dry soil. Since the compaction level of soil is a significant parameter, three soil compaction levels are considered: low, medium, and high compaction with various relative densities of 26.47, 70.58 and 91.17 % respectively. For these set of experiments, the corresponding air flow rate, inlet ambient air temperature and soil temperature were fixed at 0.006 m^3/s, 50 °C and 25 °C, respectively.

It can be noticed that the soil moisture content highly affects the thermal performance of the examined BEAHE. The temperature drop across the considered pre-cooling system increased when the soil moisture content level increased. It is also observed that the thermal performance of the system is clearly improved by increasing the level of soil relative density. As results emphasize, burring the pipes of the investigated heat exchanger in wet soil raises the effectiveness of the BEAHE. Furthermore, burring the pipes of the heat exchanger in a highly compacted soil increases the effectiveness of the BEAHE thermal performance. Accordingly, for the case of induced air with an inlet temperature of 50 °C and air flow rate of 0.006 m^3/s, an 18.29 %, 20.34 % and 22.49 % improve in the temperature drop is recorded for soil moisture content 9.87, 21.3 and 30% respectively, for high compacted soil at relative density of 91.17 %. For the case of medium compacted soil with a relative density of 70.58 % and low compacted soil with a relative density of 26.47 % an increase in the temperature drop 15.64, 16.13, and 17.81 % and 13.53, 14.16 and 16.06 % are achieved for soil moisture content 9.87, 21.3 and 30% respectively. On an average daily basis, the temperature drop across the BEAHE increases by 18.29, 20.34 and 22.49 %, when the soil moisture content reached a value of 9.87, 21.3, and 30.0 % respectively, compared to dry soil for high compaction level with soil relative density of 91.17 %.

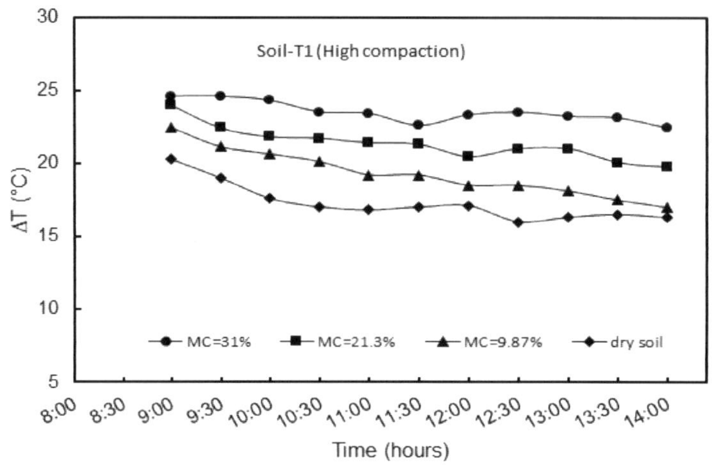

Figure 3. Effect of soil moisture content on the cooling load at soil relative density of 91.17% (high compaction).

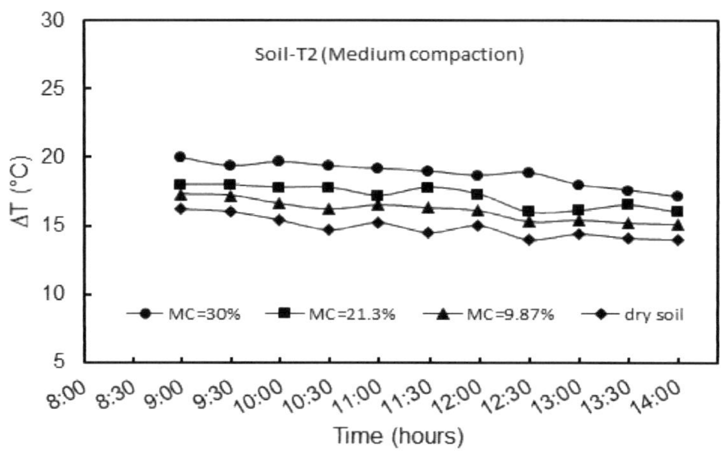

Figure 4. Effect of soil moisture content on the cooling load at soil relative density of 70.58 % (medium compaction).

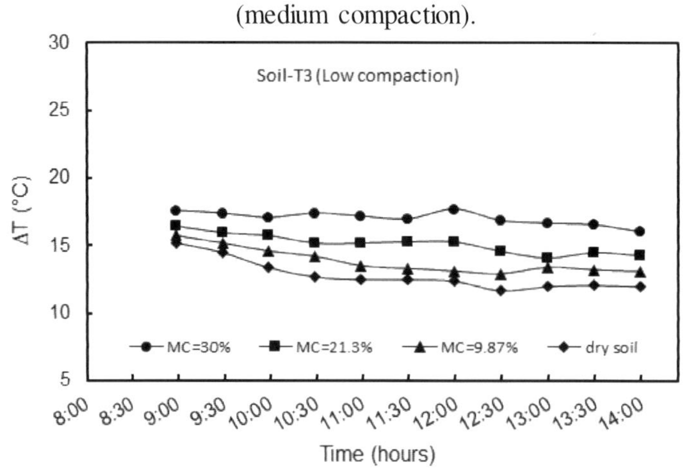

Figure 5. Effect of soil moisture content on the cooling load at soil relative density of 26.47% (low compaction).

Table 3 presents the average daily improvements percentage in the cooling loads (Q_c%) across the BEAHE for wet soil with different moisture levels and compaction levels compared with dry soil. The cooling load is enhanced by 11.74, 23.67 and 36.12% at high soil compaction (DR% = 91.17), by 9.27, 18.87 and 28.39 % at medium compaction (DR% = 70.58), and by 8.01, 16.29 and 25.83 % at low compaction (DR% = 26.47) when the moisture content is raised to 9.87, 21.3 and 30.0 % respectively, compared to dry soil.

Table 3. The daily average enhancement of the cooling load across the BEAHE.

Soil moisture contents (%)	Cooling load across the BEAHE and % enhancement compared to dry soil					
	DR% = 26.47		DR% = 70.58		DR% = 91.17	
	Q_c (W)	Q_c%	Q_c (W)	Q_c%	Q_c (W)	Q_c%
dry soil	96.14	--	111.48	--	129.47	--
9.87	103.84	8.01	121.82	9.27	144.68	11.74
21.3	111.73	16.29	132.52	18.87	160.12	23.67
30.0	120.98	25.83	143.14	28.39	176.23	36.12

The temperature drop across the BEAHE noticeably increases with increasing the soil moisture content of the soil around the considered heat exchanger. One possible explanation for this behavior is that the thermal conductivity of granular soils was observed to increase rapidly with an increase in moisture content due to the increase of replacing air porosity with water. Indeed, the dependence of thermal conductivity on moisture content has been documented by many soil science and engineering laboratory studies. This significantly increases the thermal conductivity of granular soil and the energy fluxes becomes greater since the thermal conductivity of water is 25 times larger than air. Therefore, the current results clearly demonstrate that always a high soil moisture content and soil relative density are recommended for realizing better pre-cooling effects from the BEAHE system. Furthermore, the high cooling capacity is achieved for highly moisturized compacted soil, compared to the soil with medium and low moisturized compaction levels. This is because a highly moisturized compacted soil conducts more heat due to low porosity/voids and therefore, results in high induced air temperature drop (ΔT) across the BEAHE which consequently increases the cooling capacity of the system.

3.2. Effect of soil moisture content and soil compaction on the BEAHE effectiveness

In this section, the effectiveness of the considered BEAHE system is determined for various levels of soil moisture content and soil compaction. Figures 6 through 8 illustrate the effect of soil moisture content and soil compaction on the effectiveness of BEAHE. The effectiveness of the studied BEAHE is clearly improved by increasing both soil moisture content and compaction. The daily improvement percentage in the BEAHE effectiveness when the system buried in soil with moisture content level of 9.87 %, 21.3 %, and 30 % compared to dry soil are: 29.84 % for high compaction, 20.28 % for medium compaction, and 13.20 % for low compaction. Furthermore, system effectiveness is increased by 33.34 % when the soil compaction is increased from 26.47 % to 91.17 %.

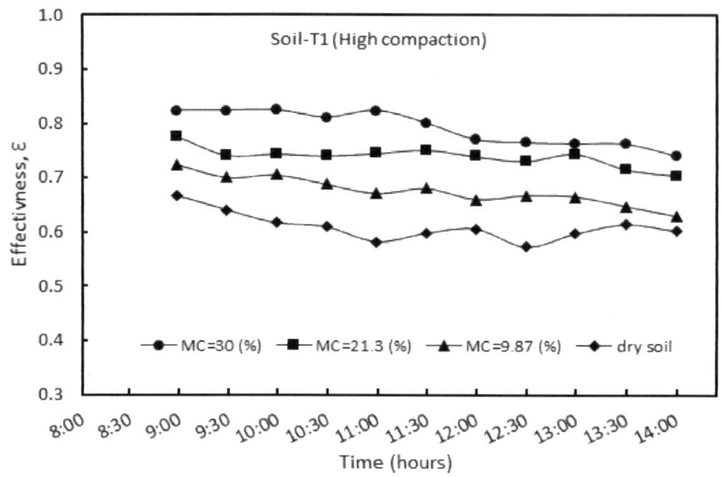

Figure 6. Effect of soil moisture content on the BEAHE effectiveness for high compaction soil.

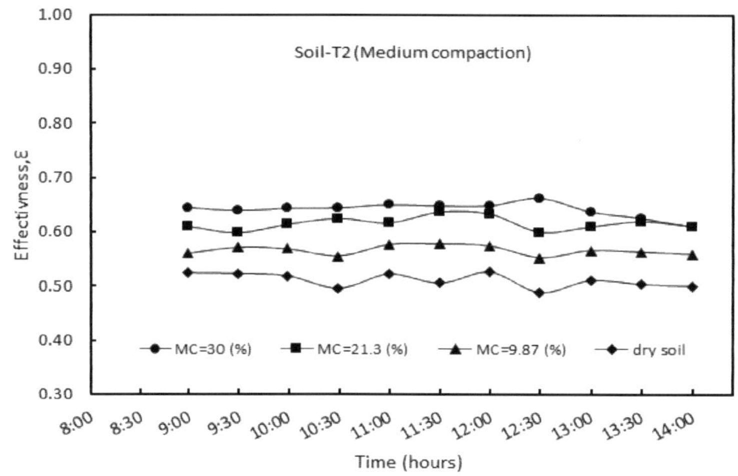

Figure 7. Effect of soil moisture content on the BEAHE effectiveness for medium compaction soil.

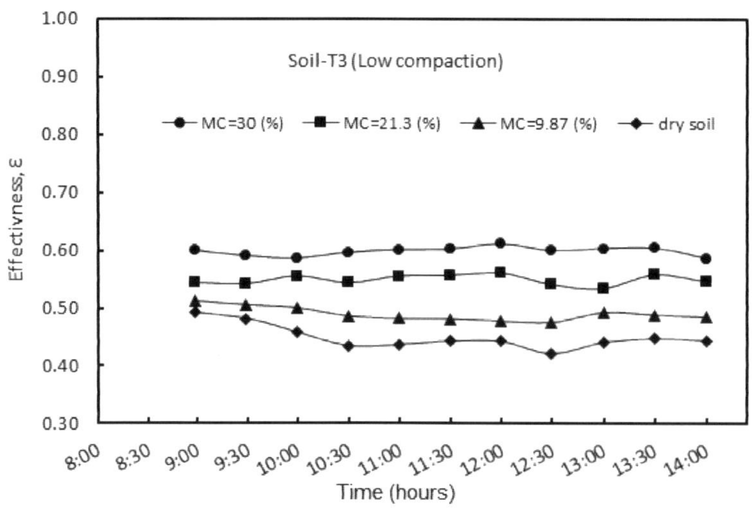

Figure 8. Effect of soil moisture content on the BEAHE effectiveness for low compaction soil.

IOP Conf. Series: Earth and Environmental Science **342** (2019) 012022 doi:10.1088/1755-1315/342/1/012022

From the above set of results, it can be observed that the effectiveness of the studied BEAHE is highly improved when its pipe buried in a dense soil with high moisture content. Since relatively dense soil with high moisture content is always associated with a low air gap and consequently high wet solid skeleton. Accordingly, for this case, the surrounding moisturized soil thermal of conductivity is more likely to raise and consequently offer a high rate of heat extraction by conduction between the pipe surface and the soil as well as between the subsequent layers of the surrounding soil.

4. Conclusions

The following can be concluded from this experimental investigation:

1. On an average daily basis, a significant induced air temperature drop of 23.5 °C is achieved at the soil moisture content of 30 % and high soil compaction of 91.17 %.
2. The induced air temperature drop across BEAHE is improved by about 18.29 %, 20.34 % and 22.49 % for soil moisture content of 9.87, 21.3 and 30 %, respectively for high compacted soil at a relative density of 91.17 %.
3. On an average daily basis, the temperature drop across the BEAHE increases by 18.29, 20.34 and 22.49 when the soil moisture content reached the value of 9.87, 21.3, and 30 % as compared to dry soil for high compaction level with soil relative density of 91.17 %.
4. The cooling load is enhanced by 11.74, 23.67 and 36.12% at high soil compaction (DR % = 91.17), by 9.27, 18.87 and 28.39 % at medium compaction (DR % = 70.58), by 8.01, 16.29 and 25.83 % at low compaction (DR % = 26.47) when the moisture content is raised to 9.87, 21.3 and 30% respectively as compared to dry soil.
5. The daily average improvement of BEAHE system effectiveness when the system buried in soil with moisture content level of 9.87%, 21.3%, and 30% compared to dry soil is 29.84% for high compaction, 20.28% for medium compaction, and 13.20% for low compaction.
6. Furthermore, the effectiveness of considered BEAHE is increased by about 33.34% when the soil compaction is increased from 26.47% to 91.17%.
7. The maximum system effectiveness of 0.823, 0.651 and 0.581 corresponding soil compaction of 91.17, 70.58 and 26.47% respectively, are achieved at a soil moisture content level of 30%.

5. References

[1] Maulbetsch J and DiFilippo M 2006, Cost and Value of Water Use at Combined-Cycle Power Plants, California Energy Commission, *PIER Energy-Related Environmental Research, CEC*-500-2006-034.

[2] Weifeng H, Dong H, Chen Y, Wenhao P, and Yiping D 2014, Mechanism of the air temperature rise at the forced draught fan inlets in an air-cooled steam condenser, *Appl. Therm. Eng.* 71, 355-63.

[3] Soni S, Pandey M, and Bartaria V 2015, Ground coupled heat exchangers: a review and applications, *Renew. Sustain. Energy Rev.* 47, 83–92.

[4] Fard A, M H, Gholami A, and Khojastehpour M 2011, Evaluation of an earth-to-air heat exchanger for the north-east of Iran with semi-arid climate. *International Journal of Green Energy,* 8(4), 499-510.

[5] Santamouris M, et al. 1995, Use of buried pipes for energy conservation in cooling of agricultural greenhouses. Solar Energy, 55(2), 111-124.

[6] Bhardwaj S, and Bansal N 1981, Temperature distribution inside ground for various surface conditions, *Building and Environment*, 10 (3) 183–192.

[7] Jacovides C, Mihalakakou G, Santamouris M, and Lewis J, On the ground temperature profile for passive cooling applications in building, *Solar Energy* 57 (3), (1996) 167–175.

[8] Bisoniya T, Kumar A, and Baredar P 2014, Cooling potential evaluation of earth–air heat exchanger system for summer season, *Int J Eng Tech Res* 2 (4), 309–316

[9] Philippe M, Bernier M, Marchio D, and Lopez S 2011, A semi-analytical model for serpentine horizontal ground heat exchangers, *HVAC & R Res.* 17 (6) 1044-1058.

[10] Congedo P, Colangelo G, and Starace G, CFD simulations of horizontal ground heat exchangers: a comparison among different configurations, *Appl. Therm. Eng.* 33e34 (2012) 24-32.

[11] Cui P, Li X, Man Y, and Fang Z 2011, Heat transfer analysis of pile geothermal heat exchangers with spiral coils, *Appl. Energy* 88 4113-4119.

[12] Park S, Sung C, Chauchois A, and Choi H 2015, Constructability and heat exchange efficiency of large diameter cast-in-place energy piles with various configurations of heat exchange pipe, *Appl. Therm. Eng.* 90, 1061-1071

[13] Naylor S, Ellett K, and Gustin A 2015, Spatiotemporal variability of ground thermal properties in glacial sediments and implications for horizontal ground heat exchanger design, *Renew. Energy* 81 21-30.

[14] Wu R, Tinjum J, and Likos W, Coupled thermal conductivity dry out curve and soil-water characteristic curve in modelling of shallow horizontal geothermal ground loops, *Geo tech Geol Eng.* 33 (2015) 193-205.

[15] Sanaye S, and Niroomand B 2010, Ground coupled heat pump: thermal-economic modeling and optimization, *Energy Convers. Manage.* 512600–2612.

[16] Wu Y, Gan G, Verhoef A, Luigi P, et al. 2010, Experimental measurement and numerical simulation of horizontal-coupled slinky ground source heat exchangers, *Appl. Therm. Eng.* 30 2574–2583.

[17] Chalhoub M, Bernier M, Coquet Y, and Philippe M 2017, A simple heat and moisture transfer model to predict ground temperature for shallow ground heat exchangers, *Renew. Energy* 103 295-307

[18] Leong W, Tarnawski V and Aittomaki A 1998, Effect of soil type and moisture content on ground heat pump performance *Int J. Refrig.* 21 (8), pp. 595–606.

[19] Piechowski M 1999, Heat and mass transfer model of a ground heat exchanger: theoretical development, *Int. J. Energy Res.* 23 571–588.

[20] Abu-Hamdeh N, Khdair A, and Reeder R 2001, A comparison of two methods used to evaluate thermal conductivity for some soil, *International J. of Heat and Mass Transfer* 44, 1073-1078.

Acknowledgment
The authors would like to acknowledge the research deanship at the Islamic University of Medina, Saudi Arabia, for providing a fund to conduct this research work.

IOP Conf. Series: Earth and Environmental Science **342** (2019) 012023 doi:10.1088/1755-1315/342/1/012023

Coordinated Scheduling of District Integrated Energy System for Enhancing Resilience in Extreme Events

S X Qi[1], X L Wang[1], Q Peng[2] and C Z Zhu[3]

[1] School of Electrical Engineering, Xi'an Jiaotong University, No.28, Xianning West Road, Xi'an, Shaanxi, P.R. China
[2] School of Electrical Engineering, Dalian University of Technology, No.2 Linggong Road, Dalian, Liaoning, P.R. China
[3] State Grid Zhejiang Electric Power Company. No. 8, Huanglong Road, Hangzhou, Zhejiang, P.R. China

Email: qishixiong@stu.xjtu.edu.cn

Abstract. In this paper, a coordinated scheduling model is proposed, which provides resilience enhancement for the district integrated energy system (DIES) against extreme events. The innovation of this model is to divide scheduling into three stages. In the normal scheduling, the objective is to minimize the operation costs until a disaster alert occurs. In the proactive scheduling, the objective is to maximize the stored energy for supplying post-disaster load demand. In the resilient scheduling, the objective is to minimize the amount of load shedding. Moreover, the time of disaster alert, disaster disruption and system recovery are uncertain. Hence, according to the intensity of disasters, the results of the proposed coordinated scheduling are analysed by three disaster scenarios. The scheduling is established by optimizing the model, which is transformed into a mixed integer linear programming. Finally, it is verified that the coordinated scheduling model proposed can effectively enhance resilience of DIES by case studies.

1. Introduction

Presently, most electricity distribution systems are designed and maintained for normal operation conditions, which cannot withstand disruptions caused by extreme events [1]. For example, in 2012, Hurricane Sandy hit the eastern shore of the United States and approximately 7.5 million customers suffered from power outages across 15 states and Washington D.C [2]. Moreover, the interaction between electricity distribution system, natural gas distribution system and district heating system is intensified based on energy hub (EH) [3]. Hence, energy industry has been more focusing on the district integrated energy system (DIES), which can maintain the loads across a small area in island mode by utilizing local distributed generators (DGs) and energy storages (ESs). The DIES is a viable solution to address the challenge of energy system resilience, when the extreme events break out. For example, the Great East Japan Earthquake caused great damage to power grid in March 2011, but the Sendai microgrid succeed to supply power to unserved loads by DGs and ESs [4]. From the above, the objective of this paper is to propose a coordinated scheduling, which can efficiently improve the energy resilience after extreme events.

The concept of resilience is different from that of reliability. The reliability evaluation focuses on high-probability and low-impact events, while the resilience analysis focuses on low-probability and high-impact events. In the literature, at least dozens of definitions for resilience can be found in different references [5]. Although the definitions of resilience are different, the connotation is basically the same. The resilience can be described as the ability to anticipate, withstand, absorb and recover quickly from disruptions caused by extreme natural catastrophes such as earthquakes, hurricane and flooding [6]. As mentioned above, the DGs and ESs in the microgrids (MGs) are the efficient back-up resources. The study in [7] proposed a smart self-healing optimization strategy, which considers several factors, including the available power supply, system configuration and load management. What's more, the MGs are managed by supporting and interchanging electricity with each other in the proposed infrastructure of the networked MGs [8]. The study in [9] presented mesh grid approach for analysing the resilience, and the result showed that MGs can improve the distribution system resilience.

The DIES is an extension of the MG, which is coupled with other energy types, such as natural gas, heat and cool. In terms of resilience, the transmission lines in power system are exposed to natural disasters, so they are quite vulnerable. However, the pipelines in natural gas system are generally underground, so they are unaffected by natural disasters, especially the surface natural disasters. Hence, it is necessary to study how to improve the resilience of the DIES. In [10], because the overhead power grid can be hardened by replacing segments of electric power grid with underground natural gas pipelines, the integrated planning is proposed for improving the power grid resilience. The study in [11] presented an approach to determine and reinforce the vulnerable components in the electricity and natural gas networks of island MG to enhance resilience against deliberate disruption. The objective of this paper is to propose a coordinated scheduling for enhancing DIES resilience in extreme events, which is divided into three operation modes.

The reminder of this paper is organized as follows: Section 2 describes the characteristics and problems of coordinated scheduling in DIES and formulates the optimization model of coordinated scheduling, which is a mixed integer linear programming (MILP). Section 3 provides numerical results to validate the effectiveness of improving resilience of the proposed coordinated scheduling. Section 4 concludes this paper.

2. Model Formulation

As noted earlier, a resilient DIES should be able to properly cope with low probability and high impact disturbances. It can be described from anticipation, absorption and restoration. Hence, the operation of DIES is divided into normal operation mode, proactive operation mode and resilient operation mode.

In the normal operation mode, the source is from purchased energy, which supplies the power loads, natural gas loads and heat loads. What's more, ESs are charged or discharged to meet the needs of consumers. The normal operation model is formulated by:

Objective function:

$$C^{\text{nor}} = \min \sum_{t=1}^{T_1} \left(C_t^{\text{s}} + C_t^{\text{sto}} + C_t^{\text{w}} + C_t^{\text{shed}} \right) \tag{1}$$

$$C_t^{\text{s}} = P_t^{\text{s}} \lambda_t^{\text{e}} + F_t^{\text{s}} \lambda_t^{\text{g}} \tag{2}$$

$$C_t^{\text{sto}} = \sum_{i \in G^{\text{b}}} \left(P_{i,t}^{\text{in}} \lambda_{i,t}^{\text{b,in}} + P_{i,t}^{\text{out}} \lambda_{i,t}^{\text{b,out}} \right) + \sum_{i \in G^{\text{g}}} \left(F_{i,t}^{\text{in}} \lambda_{i,t}^{\text{g,in}} + F_{i,t}^{\text{out}} \lambda_{i,t}^{\text{g,out}} \right) + \sum_{i \in G^{\text{h}}} \left(Q_{i,t}^{\text{in}} \lambda_{i,t}^{\text{h,in}} + Q_{i,t}^{\text{out}} \lambda_{i,t}^{\text{h,out}} \right) \tag{3}$$

$$C_t^{\text{w}} = \pi^{\text{w}} P_t^{\text{w,c}} \tag{4}$$

$$C_t^{\text{shed}} = \pi^{\text{e}} P_t^{\text{shed}} + \pi^{\text{g}} F_t^{\text{shed}} + \pi^{\text{h}} Q_t^{\text{shed}} \tag{5}$$

Subject to:

$$P_t^{\text{s}} + P_t^{\text{w}} + \sum_{i \in G^{\text{CHP}}} P_{i,t}^{\text{CHP}} + \sum_{i \in G^{\text{b}}} P_{i,t}^{\text{out}} = P_t^{\text{d}} + \sum_{i \in G^{\text{pump}}} P_{i,t}^{\text{pump}} + \sum_{i \in G^{\text{b}}} P_{i,t}^{\text{in}} - P_t^{\text{shed}} \quad \forall t \tag{6}$$

$$F_t^{\text{s}} + \sum_{i \in G^{\text{g}}} F_{i,t}^{\text{out}} = F_t^{\text{d}} + \sum_{i \in G^{\text{CHP}}} F_{i,t}^{\text{CHP}} + \sum_{i \in G^{\text{boiler}}} F_{i,t}^{\text{boiler}} + \sum_{i \in G^{\text{g}}} F_{i,t}^{\text{in}} - F_t^{\text{shed}} \quad \forall t \tag{7}$$

$$\sum_{i \in G^{\text{CHP}}} Q_{i,t}^{\text{CHP}} + \sum_{i \in G^{\text{boiler}}} Q_{i,t}^{\text{boiler}} + \sum_{i \in G^{\text{pump}}} Q_{i,t}^{\text{pump}} + \sum_{i \in G^{\text{h}}} Q_{i,t}^{\text{out}} = Q_t^{\text{d}} + \sum_{i \in G^{\text{h}}} Q_{i,t}^{\text{in}} - Q_t^{\text{shed}} \quad \forall t \tag{8}$$

$$0 \le P_t^{s} \le \bar{P}_t^{s} \qquad \forall t \tag{9}$$

$$0 \le F_t^{s} \le \bar{F}_t^{s} \qquad \forall t \tag{10}$$

$$P_{i,t}^{CHP} = \partial_i \eta_i^{e,CHP} F_{i,t}^{CHP} \qquad \forall i, \forall t \tag{11}$$

$$Q_{i,t}^{CHP} = (1-\partial_i) \eta_i^{h,CHP} F_{i,t}^{CHP} \qquad \forall i, \forall t \tag{12}$$

$$0 \le P_{i,t}^{CHP} \le \bar{P}_i^{CHP} \qquad \forall i, \forall t \tag{13}$$

$$-\Delta \underline{P}_i^{CHP} \le P_{i,t+1}^{CHP} - P_{i,t}^{CHP} \le \Delta \bar{P}_i^{CHP} \qquad \forall i, \forall t \tag{14}$$

$$Q_{i,t}^{pump} = \eta_i^{pump} P_{i,t}^{pump} \qquad \forall i, \forall t \tag{15}$$

$$0 \le Q_{i,t}^{pump} \le \bar{Q}_i^{pump} \qquad \forall i, \forall t \tag{16}$$

$$-\Delta \underline{Q}_i^{pump} \le Q_{i,t+1}^{pump} - Q_{i,t}^{pump} \le \Delta \bar{Q}_i^{pump} \qquad \forall i, \forall t \tag{17}$$

$$Q_{i,t}^{boiler} = \eta_i^{boiler} P_{i,t}^{boiler} \qquad \forall i, \forall t \tag{18}$$

$$0 \le Q_{i,t}^{boiler} \le \bar{Q}_i^{boiler} \qquad \forall i, \forall t \tag{19}$$

$$-\Delta \underline{Q}_i^{boiler} \le Q_{i,t+1}^{boiler} - Q_{i,t}^{boiler} \le \Delta \bar{Q}_i^{boiler} \qquad \forall i, \forall t \tag{20}$$

$$E_{i,t+1}^{x} = E_{i,t}^{x} + X_{i,t}^{in} \eta_i^{x,in} \Delta t - \frac{X_{i,t}^{out} \Delta t}{\eta_i^{x,out}} \qquad \forall i, \forall t \tag{21}$$

$$0 \le E_{i,t}^{x} \le \bar{E}_i^{x} \qquad \forall i, \forall t \tag{22}$$

$$0 \le X_{i,t}^{in} \le \beta_{i,t}^{x} \bar{X}_i^{in} \qquad \forall i, \forall t \tag{23}$$

$$0 \le X_{i,t}^{out} \le (1-\beta_{i,t}^{x}) \bar{X}_i^{out} \qquad \forall i, \forall t \tag{24}$$

$$E_{i,T}^{x} = E_{i,0}^{x} \tag{25}$$

$$0 \le P_t^{w} \le \bar{P}_t^{w} \qquad \forall t \tag{26}$$

$$P_t^{w,c} = \bar{P}_t^{w} - P_t^{w} \qquad \forall t \tag{27}$$

$$0 \le P_t^{shed} \le P_t^{d} \qquad \forall t \tag{28}$$

$$0 \le F_t^{shed} \le F_t^{d} \qquad \forall t \tag{29}$$

$$0 \le Q_t^{shed} \le Q_t^{d} \qquad \forall t \tag{30}$$

The objective of the scheduling is to minimize the operation costs C^{nor} including energy purchasing cost C_t^{s}, storage device operation cost C_t^{sto}, wind curtailment cost C_t^{w} and load shedding cost C_t^{shed}. Where P_t^{s} and F_t^{s} are purchased electricity and gas; λ_t^{e} and λ_t^{g} are electricity price and gas price, respectively; G^{b}, G^{g} and G^{h} are sets of batteries, gas storages and heat storages; $P_{i,t}^{in}$, $P_{i,t}^{out}$, $F_{i,t}^{in}$, $F_{i,t}^{out}$, $Q_{i,t}^{in}$ and $Q_{i,t}^{out}$ are amount of charge and discharge of storage devices; $\lambda_{i,t}^{b,in}$, $\lambda_{i,t}^{b,out}$, $\lambda_{i,t}^{g,in}$, $\lambda_{i,t}^{g,out}$, $\lambda_{i,t}^{h,in}$ and $\lambda_{i,t}^{h,out}$ are operation cost of storage devices; $P_t^{w,c}$ is power of wind abandonment; π^{w} is cost coefficient of wind abandonment; P_t^{shed}, F_t^{shed} and Q_t^{shed} are load shedding of power, gas and heat, respectively; π^{e}, π^{g} and π^{h} are cost coefficients of load shedding.

Where constraints (6)-(8) are energy balance constraints; constraints (9)-(10) restrict purchased energy; constraints (11)-(20) describes operation of CHP units, pumps and boilers; constraints (11)-(12), (15), (18) are energy conversion constraints, constraints (13), (16), (19) enforce the upper output limits of CHP units, pumps and boilers; constraints (14), (17), (20) limit ramp-up and ramp-down of CHP units, pumps and boilers; constraints (21) show the temporal continuity of ESs; constraints (22) refer to storage capacity limit; constraints (23)-(24) guarantee the states of charge and discharge are at a different time; constraints (25) guarantee balance of charge and discharge; constraints (26) limit output of wind power; constraints (27) show wind curtailment; constraints (28)-(29) limit the load shedding amount.

Where P_t^{d}, F_t^{d} and Q_t^{d} are power loads, gas loads and heat loads; G^{CHP}, G^{boiler} and G^{pump} are sets of combined heat and power (CHP) units, boilers and pumps; $P_{i,t}^{\mathrm{CHP}}$, $Q_{i,t}^{\mathrm{CHP}}$, $Q_{i,t}^{\mathrm{boiler}}$ and $Q_{i,t}^{\mathrm{pump}}$ are outputs of CHP units, boilers and pumps; $F_{i,t}^{\mathrm{CHP}}$, $F_{i,t}^{\mathrm{boiler}}$ and $P_{i,t}^{\mathrm{pump}}$ are gas consumption of CHP units and boilers and power consumption of pumps, respectively; $\bar{P}_t^{\mathrm{s}}$ and $\bar{F}_t^{\mathrm{s}}$ are maximum purchased energy; ∂_i is ratio coefficient of electricity and heat; $\eta_i^{\mathrm{e,CHP}}$, $\eta_i^{\mathrm{h,CHP}}$, η_i^{pump} and η_i^{boiler} are efficiency of energy conversion; $\bar{P}_i^{\mathrm{CHP}}$, $\bar{Q}_i^{\mathrm{pump}}$ and $\bar{Q}_i^{\mathrm{boiler}}$ are maximum outputs of CHP units, pumps and boilers; $\Delta\bar{P}_i^{\mathrm{CHP}}$, $\Delta\bar{Q}_i^{\mathrm{pump}}$, $\Delta\bar{Q}_i^{\mathrm{boiler}}$, $\Delta\underline{P}_i^{\mathrm{CHP}}$, $\Delta\underline{Q}_i^{\mathrm{pump}}$ and $\Delta\underline{Q}_i^{\mathrm{boiler}}$ are maximum ramp-up and ramp-down of CHP units, pumps and boilers; $E_{i,t}^{\mathrm{x}}$ is amount of ESs; $\eta_i^{\mathrm{x,in}}$ and $\eta_i^{\mathrm{x,out}}$ are efficiency of charge and discharge; $X_{i,t}^{\mathrm{in}}$ and $X_{i,t}^{\mathrm{out}}$ are charge and discharge power of ESs; $\bar{X}_i^{\mathrm{in}}$ and $\bar{X}_i^{\mathrm{out}}$ are maximum of charge and discharge power; $\bar{E}_i^{\mathrm{x}}$ is storage capacity; $\beta_{i,t}^{\mathrm{x}}$ is a binary variable, which guarantees the states of charge and discharge are at a different time; x represents type of energy, including electricity, gas and heat; $\bar{P}_t^{\mathrm{w}}$ is maximum output of wind generator.

Assume that disaster alert is received at time T_1. After T_1, DIES will enter the proactive operation mode. In this mode, the source is from purchased energy, not only to supply the power loads, natural gas loads and heat loads, but also to charge ESs as much as possible. The proactive operation model is formulated by:

Objective function:

$$C^{\mathrm{pro}} = \min \sum_{t=T_1+1}^{T_2} \left(C_t^{\mathrm{s}} + C_t^{\mathrm{sto}} + C_t^{\mathrm{w}} + C_t^{\mathrm{shed}} + C_t^{\mathrm{emp}} \right) \tag{31}$$

$$C_t^{\mathrm{emp}} = \pi^{\mathrm{b}}(\bar{E}_{i,t}^{\mathrm{b}} - E_{i,t}^{\mathrm{b}}) + \pi^{\mathrm{g}}(\bar{E}_{i,t}^{\mathrm{g}} - E_{i,t}^{\mathrm{g}}) + \pi^{\mathrm{h}}(\bar{E}_{i,t}^{\mathrm{h}} - E_{i,t}^{\mathrm{h}}) \tag{32}$$

Subject to:

Constraints (6)-(24), (26)-(30).

Where π^{b}, π^{g} and π^{h} are penalty coefficients of unused storage amount. In the case where purchased energy is allowed, whether ESs can be filled depends on the actual occurrence time T_2 of disaster. When disaster occurs at time T_2, the purchased energy will be limited. Hence, DIES has to optimize the stored energy in the proactive operation mode so that load shedding can be reduced as much as possible. The resilient operation model is formulated by:

Objective function:

$$C^{\mathrm{res}} = \min \sum_{t=T_2+1}^{T_3} \left(C_t^{\mathrm{sto}} + C_t^{\mathrm{w}} + C_t^{\mathrm{shed}} \right) \tag{33}$$

Subject to:

$$0 \le P_t^{\mathrm{s}} \le \varepsilon_t^{\mathrm{e}} \bar{P}_t^{\mathrm{s}} \quad \forall t \tag{34}$$

$$0 \le F_t^{\mathrm{s}} \le \varepsilon_t^{\mathrm{g}} \bar{F}_t^{\mathrm{s}} \quad \forall t \tag{35}$$

Constraints (6)-(8), (11)-(24), (26)-(30).

Where $\varepsilon_t^{\mathrm{e}}$ and $\varepsilon_t^{\mathrm{g}}$ are recovery coefficients of electrical power system and natural gas system. The recovery coefficients are between 0-1. It is worth mentioning that the cost coefficients of load shedding (π^{e}, π^{g}, π^{h}) will gradually decrease with the increase of scheduling period, which encourages to supply load in the first few periods after disaster, thus avoiding multiple solutions. The reason for setting cost coefficients of load shedding is that the failures of power distribution network and gas distribution network are being gradually repaired. According to the intensity of disasters, if the distribution network has been repaired and purchased energy has been allowed before the stored energy is exhausted, load shedding will be avoided.

The proposed coordinated scheduling model of DIES is a MILP, which can be solved by commercial optimization solver, such as Cplex and Gurobi.

3. Numerical Results

In this section, three scenarios in which integrated energy system is damaged by disaster are simulated. On this basis, the proposed coordinated scheduling model is examined on a DIES in mild, moderate, severe damage scenarios, respectively. Through different disaster scenarios, the enhancement effect on DIES resilience of the scheduling model proposed is analysed.

In case studies, the DIES includes two CHP units, two pumps, one boiler, two batteries, one gas storage and one heat storage. What's more, the scheduling horizon is set to be 24 hours. It is assumed that the controller of DIES receives a disaster alert at the 5th hour, and the disaster attacks the DIES at 15th. Based on this assumption, normal operation, proactive operation and resilient operation cover 4 hours, 10 hours and 10 hours, respectively. The purchased energy in different scenarios is shown in figure 1.

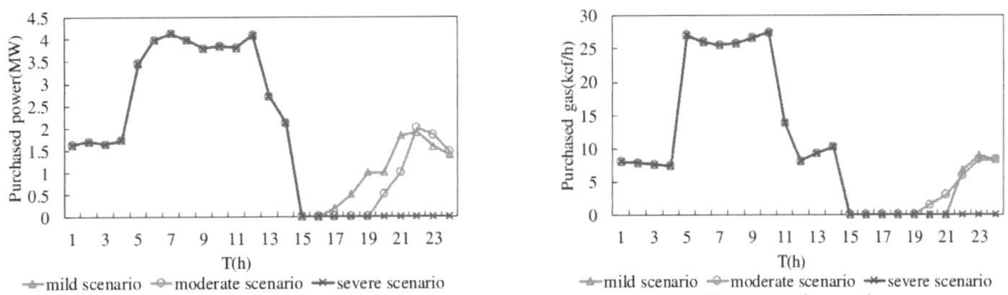

(a) The purchased power (b) The purchased gas

Figure 1. The purchased energy in different scenarios

As shown in figure 1, the 1 to 4 hours is during the normal operation and the purchased energy matches load demand. Next, the 5 to 14 hours is during the proactive operation, in which the purchased energy is extremely increased for load demand and ESs. And then, the 15 to 24 hours is the resilient operation. In the mild scenario and moderate scenario, the purchased energy is gradually allowed to a lager extent. However, it is totally limited in the severe scenario. Whether the purchased energy is allowed to be related to whether load is supplied. The load supply is shown in figure 2, figure 3 and figure 4.

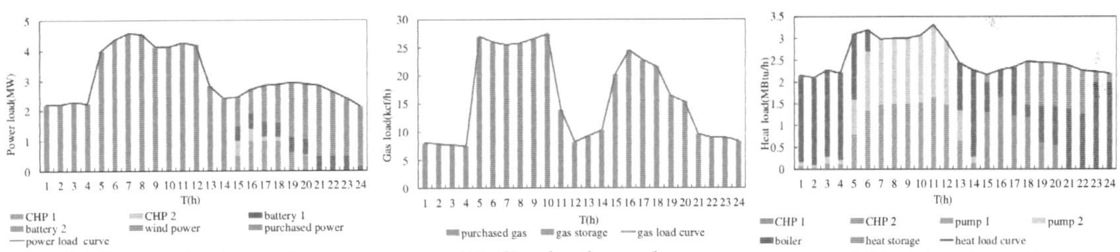

(a) Power load supply (b) Gas load supply (c) Heat load supply

Figure 2. The load supply in mild scenario

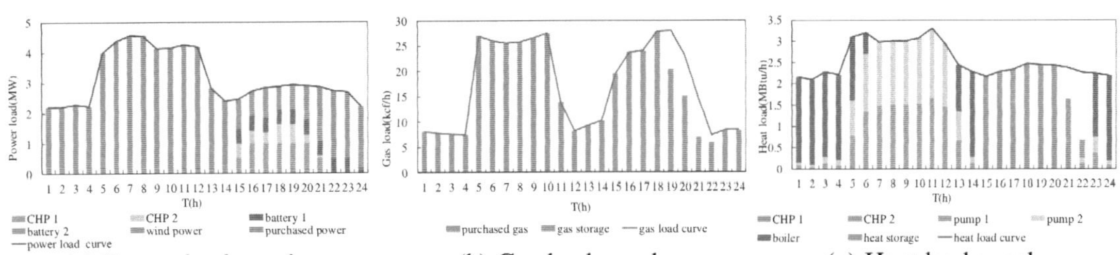

(a) Power load supply (b) Gas load supply (c) Heat load supply

Figure 3. The load supply in moderate scenario

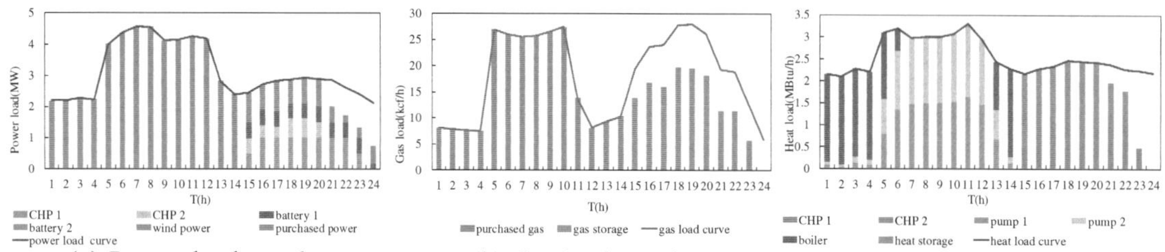

(a) Power load supply (b) Gas load supply (c) Heat load supply

Figure 4. The load supply in severe scenario

It is worth explaining that the power load curves and gas load curves are various in different scenarios, which is caused by electricity-gas-heat coupling equipment. For example, although matching the same heat load curve, it is possible that CHP units and boilers consume natural gas to generate heat, or pumps consumes electric power to generate heat.

As shown in figure 2, since the purchased energy is gradually allowed after 17th hour, the stored energy during the proactive operation is sufficient to support the load after disaster, achieving the goal of withstanding disaster without load shedding, and improving the resilience of DIES. However, as shown in figure 3, the purchased energy is not allowed until 20th hour, so part of gas loads and heat loads has to be out of supply in the 19th to 22nd hours. In this paper, when the stored energy is insufficient, the supply for gas load and heat load can be terminated, while power load needs to be supplied as much as possible. Furthermore, when the stored energy is exhausted in severe scenario, DIES has to lose load more seriously in figure 4. In conclusion, it can effectively avoid load shedding in mild disaster scenario. However, as the degree of natural disasters deepens, it will inevitably lose load.

In order to verify that coordinated scheduling model proposed can effectively enhance the resilience of DIES, this paper compares the operation without proactive stage. We choose the severe scenario, and the total power load shedding, gas load shedding and heat load shedding are 18.26MW, 74.17kcf/h and 20.20MBtu/h in the original DIES. However, in figure 4, the total load shedding is 4.26MW, 72.55kcf/h and 4.83MBtu/h, respectively, where we can see a significant decline both in power load and heat load with a gentle drop in gas load. This shows that coordinated scheduling model proposed can improve the resilience of DIES.

4. Conclusion

This paper studies the resilience of DIES. A coordinated scheduling model is proposed, which can enhance resilience of DIES against extreme events. This model is divided into three steps of scheduling, which can all be converted into mixed integer linear programming. They correspond to three operation modes of DIES: the normal operation, the proactive operation and the resilient operation. Actually, it is essentially a method of scheduling ESs in advance to prepare sufficient energy for avoiding load shedding as much as possible after disasters. The case studies show that coordinated scheduling model proposed can effectively alleviate load shedding in different intensity of disaster scenario and provide resilience enhancement for DIES greatly.

5. References

[1] S. S. Ma, L. Su, Z. Y. Wang, F. Qiu, and G. Guo, "Resilience Enhancement of Distribution Grids Against Extreme Weather Events," *IEEE Trans on Power Systems*, vol. 33, no.5, pp. 4842-4853, Sep. 2018.

[2] L. Che, M. Khodayar, and M. Shahidehpour, "Only connect: Microgrids for distribution system restoration," *IEEE Power Energy Mag.*, vol. 12, no.1, pp. 70-81, Jan.-Feb. 2014.

[3] M. Geidl, G. Koeppel, P. Favre-Peerod, B. KLöckl, G. Andersson, and K. Fröhlich, "Energy hubs for the future," *IEEE Power Energy Mag.*, vol. 5, no. 1, pp. 2130, Jan.-Feb. 2007.

[4] C. Marnay, H. Aki, K. Hirose, A. Kwasinski, S, Ogura, and T. Shinji, "Japan's pivot to resilience: how two microgrids fared after the 2011 Earthquake", *IEEE Power Energy Mag.*, vol. 13, no. 3, pp. 44-57, May-Jun. 2015.

[5] M. Zare-Bahramabadi, A. Abbaspour, M. Fotuhi-Firuzabad, and M. Moeini-Aghtaie, "Resilience-based framework for switch placement problem in power distribution systems," *IET Generation, Transmission & Distribution*, vol. 12, no. 5, pp. 1229230, Jan. 2018.

[6] A. Gholami, F. Aminifar, and M. Shahidehpour, "Front lines against the darkness enhancing the resilience of the electricity grid through microgrid facilities," *IEEE Electrification Magazine*, vol. 4, no. 1, pp. 1274, Mar. 2016.

[7] M. A. Owaifeer, and M. Al-Muhaini, "MILP-based technique for smart self-healing grids," *IET Generation, Transmission & Distribution*, vol. 12, no. 10, pp. 23022316, Apr. 2018.

[8] Z. Y. Wang, B. Chen, J. H. Wang, and C. Chen, "Networked microgrids for self-healing power systems," *IEEE Trans. on Smart Grid*, vol. 7, no. 1, pp. 310-319, Jan. 2016.

[9] X. D. Liu, M. Shahidehpour, Z. Y. Li, X. Liu, Y. J. Cao, and Z. H. Bie, "Microgrids for enhancing the power grid resilience in extreme conditions," *IEEE Trans. on Smart Grid*, vol. 8, no. 2, pp. 589-597, Mar. 2017.

[10] C. C. Shao, M. Shahidehpour, X. F. Wang, X. L. Wang, and B. Y. Wang, "Integrated planning of electricity and natural gas transportation systems for enhancing the power grid resilience," *IEEE Trans on Power Systems*, vol. 32, no. 6, pp. 4418-4429, Nov. 2017.

[11] S. MANSHADI, and M. KHODAYAR, "Preventive reinforcement under uncertainty for islanded microgrids with electricity and natural gas networks," *Journal of Modern Power Systems and Clean Energy*, vol. 6, no. 6, pp. 1229233, Nov. 2018.

Acknowledgments

This work is supported in part by Science and Technology Foundation of SGCC (Research on Morphologies and Pathways of Future Power System)

IOP Conf. Series: Earth and Environmental Science **342** (2019) 012024 doi:10.1088/1755-1315/342/1/012024

Smoothing Control Strategy of Wind Power Based on Adaptive Adjustment of Fluctuation Coefficient

Luo Yi, Su Kunlin, Yin Jiye

School of Control and Computer Engineering, North China Electric Power University, Beijing 102206, China

su1kun3lin5@163.com

Abstract. Due to the strong randomness and fluctuation of wind power, a direct access to the power grid will be an adverse effect on the quality of the power. The energy storage system can not only reduce the phenomenon of abandoned wind power, but also improve the quality of the power grid. In order to smooth the fluctuation of wind power, we propose a smooth control strategy of wind power based on adaptive adjustment of fluctuation coefficient is optimized. The optimal model is designed which is based on minimize the energy storage capacity. The method of sliding window and particle swarm optimization(PSO) is used to solve the problem. The wind power grid-connected power value is obtained.

1. Introduction

With the rapid development of wind power technology, the total installed capacity of wind farms continues to increase, and the wind power penetration rate is also increasing. In order to improve the grid-connecting capacity of wind power and reduce the phenomenon of wind curtailment, it is possible to equip wind farms with a certain capacity of energy storage system[1]-[2].

How to use less energy storage capacity to stabilize the wind power fluctuations during wind power grid connection and it put forward higher requirements for control systems. In the research of control algorithms, more and more control algorithms have been proposed. The first-order low-pass filtering algorithm is widely used because of its simple principle and fast calculation speed[3]-[5].However, it has a certain delay in tracking wind power changes. In[6], only the particle swarm optimization (PSO) algorithm is considered, and the overall data impact on the grid-connected wind power is not considered when the particle swarm is trapped in the local optimal solution. In [7]-[8], a fuzzy control strategy is proposed to stabilize the wind power through 'state of charging (SOC) changes. In[9], Using the empirical mode decomposition method to quickly analyze the wind power fluctuations, we obtain the energy storage stabilization control command. Based on the above research, this paper considers the smoothing effect of wind power grid connection and the reduction of energy storage cost. Considering the actual situation, a control algorithm based on adaptive adjustment of fluctuation coefficient is designed.

2. Wind power smoothing strategy

At present, the energy storage system has different configurations according to the actual situation of the wind farm. There are mainly two structures: distributed and centralized. The distributed wind farm energy storage system is equipped with a small capacity energy storage device for each fan in the wind

IOP Conf. Series: Earth and Environmental Science **342** (2019) 012024 doi:10.1088/1755-1315/342/1/012024

farm. And the centralized wind farm energy storage system is equipped with a large capacity energy storage device for the entire wind farm. This paper selects the centralized wind farm energy storage system for analysis.

2.1 Proposal of Wind Power Smoothing Strategy

The energy storage system needs to timely absorb or release the difference between the wind power and the grid-connected power, and improving the power quality of the power grid. Energy balance on the centralized wind energy storage system:

$$P_{ess,t} = P_{w,t} - P_{o,t} \tag{1}$$

Where, $P_{ess,t}$ is the power absorbed or released by energy storage . $P_{o,t}$ is the grid-connected power for time t . $P_{w,t}$ is the total wind power of selected fans for time t . $P_{ess,t} > 0$ is that the energy storage system is charging. $P_{ess,t} < 0$ is that the energy storage system is discharging.

The slope of the wind power curve directly reflects the fluctuation of power. The greater the slope at the moment of $t \sim (t+\Delta t)$, the greater the change rate of power and the more obvious the fluctuation. Thus the wind power integration can be obtained, as shown in formula (2).

$$\begin{cases} \dfrac{P_o(t+\Delta t) - P_o(t)}{\Delta t} = w_t \times \dfrac{P_w(t+\Delta t) - p_w(t)}{\Delta t} \\ P_o(t) = (1-w_t) \times P_o(t-1) + w_t \times P_w(t) \end{cases} \tag{2}$$

Where, t is the current time and Δt represents the step size. w_t is the fluctuation coefficient. It can be known that the smaller the fluctuation coefficient , the better smoothing effect on the wind power and the more stable the grid-connected power of the wind farm. However, this may lead to excessive repression.

2.2 Wind power fluctuation definition

When wind power is connected to the power grid, the fluctuation of its active power is the main consideration of this paper. The sampling time is one minute. Define the fluctuation ratio of wind power integration within one minute:

$$\Delta P_{o,1\min}\% = | P_{o,t} - P_{o,t-1} | / P_r \tag{3}$$

Where, $P_{o,t}$ is the grid-connected power for time t . $P_{o,t-1}$ is the grid power of the previous moment. P_r is wind farm rated power.

Define the fluctuation ratio of wind power grid connected within ten minutes:

$$\Delta P_{o,10\min}\% = [\max(P_{o,t\sim t+9}) - \min(P_{o,t\sim t+9})] / P_r \tag{4}$$

Where, $\max(P_{o,t\sim t+9})$ is the maximum value of grid-connected power in ten minutes. $\min(P_{o,t\sim t+9})$ is the minimum value of grid-connected power in ten minutes.

From the definition of wind power fluctuation rate of ten minutes, the power fluctuation of ten minutes is accumulated by the power fluctuation of one minute. The wind power fluctuation constraint of one minute in the wind power grid connection should be considered first, and then the wind power fluctuation of ten minutes is considered. According to China's "Technical Rule for Connecting Wind Farm to Power System", if wind power is to be connected to the grid, the change rate of wind farm output power must be limited. Table 1 is the technical regulations for wind power fluctuations in China[10].

Table 1. Maximum variation of active power in wind farm

Wind farm installation capacity /MW	Ten minutes active power change maximum limit /MW	One minute active power change maximum limit /MW
<30	10	3

30~150	Installed capacity /3	Installed capacity /10
>150	50	15

3 Particle Swarm Optimization Algorithm Based on Sliding Window

For the economic analysis of engineering applications, it is necessary to minimize the size of the energy storage system to reduce the cost of the wind farm energy storage system. The particle swarm modeling parameter of the sliding window selects the fluctuation coefficient w_t in equation (2). In this paper, eight 1.5 MW units of a centralized wind farm were selected and analyzed using the collected one minute fan power data. The particle swarm optimization algorithm is used to optimize the fluctuation coefficient. The output of the model is the power absorbed or released by the energy storage system. Because the particle group is easy to fall into the local optimum, it has a great influence on the output. On this basis, we introduce the concept of sliding window to eliminate the impact of particle swarming into local optimum on the overall data of energy storage system. Usually the price of an energy storage system is proportional to the size of its energy storage capacity, so minimizing the capacity of the energy storage system is an objective function [11]. Define the objective function as:

$$F = \min\sqrt{\frac{1}{n}\sum_{t=1}^{n}[P_{o,t}(w_t) - P_{w,t}]^2} \tag{5}$$

Where, F is the smallest energy storage capacity. n is sampling points in the time scale. $P_{o,t}(w_t)$ is the target power after the smoothing of the fluctuation coefficient for time t.

The constraints are:

(1) The fluctuations per minute are less than 2% of the rated power of the system.

(2) The fluctuations every ten minutes are less than 20% of the rated power of the system.

$$\begin{cases} \Delta P_{o,1\min}\% \le k_1 \\ \Delta P_{o,10\min}\% \le k_2 \end{cases} \tag{6}$$

Where, $\Delta P_{o,1\min}\%$ is the volatility within one minute. $\Delta P_{o,10\min}\%$ is the volatility within ten minutes. k_1 and k_2 is Maximum power fluctuation.

In order to avoid the excessive constraints in the optimization process and avoid the influence of the defects of the PSO algorithm on the overall data of the energy storage system, we use the sliding window method (Figure2) to dynamically optimize the fluctuation coefficient. We can rewrite equation (5) as:

$$\begin{cases} P = \min\sqrt{\frac{1}{N_t}\sum_{t=i-9}^{i}[P_{o,t}(w_t) - P_{w,t}]^2} \\ F = \sum_{i=N_t}^{n}[P_{o,t}(w_t') - P_{w,t}] \end{cases} \tag{7}$$

Where, P is the smallest energy storage capacity in the set sliding window width. N_t is sliding window width. $P_{o,t}(w_t')$ is the target power obtained after the smoothing of the optimal fluctuation coefficient obtained in the sliding window. F is the smallest energy storage capacity in the total sampling point n of the system.

Since the system needs to meet the wind power grid-connected power fluctuation requirements, here we set the sliding window width N_t to 10.

Obtain continuous $N_t - 1$ wind power data before t time and t time. Establish an optimization model equations (2) and (7) with the variable w_t. The constraint is equation (6). The PSO algorithm is used to obtain the optimal fluctuation coefficient w_t', which is used as the fluctuation coefficient at time t, so as to obtain the minimum compensation power of the energy storage system.

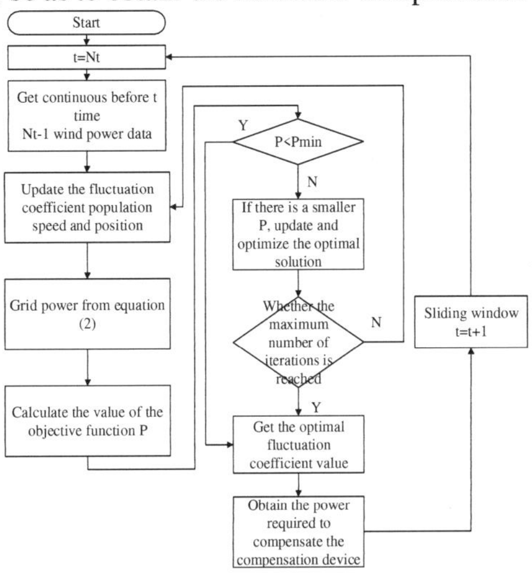

Figure 1. Flow chart of particle swarm algorithm based on sliding window

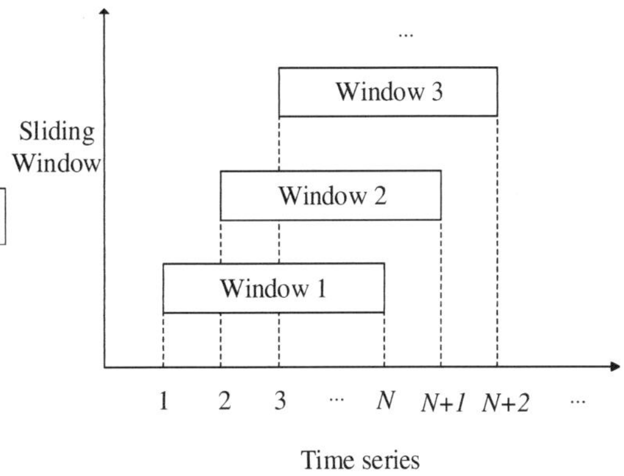

Figure 2. Slide window diagram

4 Energy storage device capacity calculation

The battery is widely used in all walks of life. It has the advantages of stable voltage, high safety and flexible energy storage capacity. Based on the above advantages, this paper uses the battery as an energy storage unit to stabilize wind power fluctuations [12]-[13].

4.1 Rated power

Considering the charging and discharging efficiency of the system:

$$P_{ess,t} = \begin{cases} P_{ess,t} / m_1, P_{ess,t} > 0 \\ P_{ess,t} / m_2, P_{ess,t} < 0 \end{cases}, t = 1,2,....n \tag{8}$$

Where, $P_{ess,t}$ is the storage energy power compensation at time t. m_1 is the charging efficiency of energy storage battery. m_2 is the discharge efficiency of energy storage battery. n is the sample points.

During the entire sampling period, the rated power of energy storage is defined as the maximum absolute value of the power compensation required by the energy storage system:

$$P_e = \max\{| P_{ess} |\} \tag{9}$$

4.2 Nominal capacity

Charge and discharge of energy storage at different times relative to the initial time:

$$E(t) = E(0) + \sum_{0}^{t} P_{ess,t} .t, t = 1,2...n \tag{10}$$

Where, t is the sampling interval, n is the number of sampling points. $E(0)$ is the initial energy of the energy storage battery. Here we assume that the initial energy is 0 for simplicity.

In practical applications, energy storage devices generally have problems of overcharging and overdischarging. Overcharge and overdischarge will affect the life of the energy storage system. Therefore, the battery will have a certain working range, and state of charging (SOC) is usually between 0.2 and 0.8. The required rated capacity of the battery:

$$C = (\max_{t=1\sim n} | E(t) |) / 0.5 \tag{11}$$

5 Examples Analysis

The actual output power of a certain wind farm is analyzed in China in 2015. As shown in the flowchart of Figure 1, the control strategy combining sliding window with PSO is used to solve the problem.

The actual output grid-connected power of the wind farm is obtained by this paper's algorithm and the first-order low-pass filter algorithm as shown in Figure 3. By comparison, it can be found that the wind power grid-connected power is smoothed by the control algorithm of this paper, which can reduce the wind power fluctuation and make the power curve smoother. The local data of the wind farm in Figure 3 is selected for analysis. As shown in Figure 4.

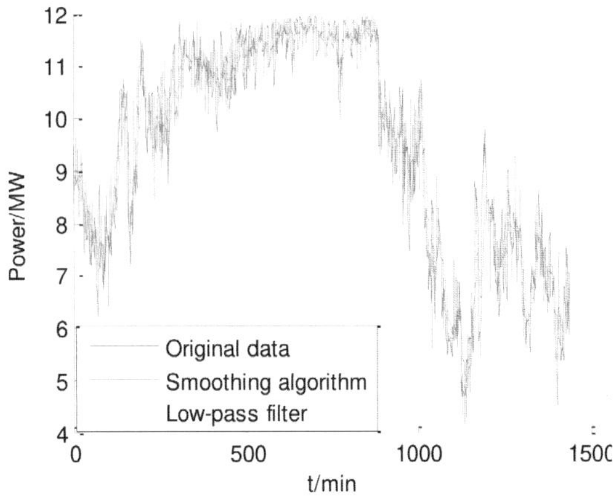

Figure 3. Comparison of wind power smoothing effect at 1440 min

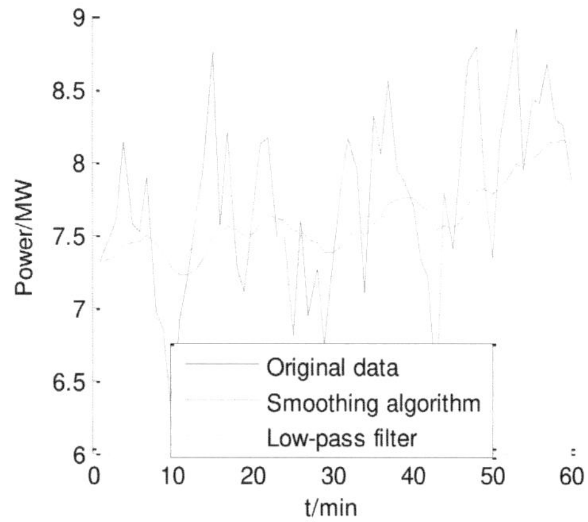

Figure 4. Comparison of wind power smoothing effect at 60 min

Figure 5 and Figure 6 are the power fluctuations of one minute and ten minutes for grid-connected power in Figure 3. It can be found that the volatility of the particle swarm optimization algorithm based on sliding window is significantly lower than that of the first-order low-pass filtering algorithm. The feasibility and effectiveness of the control strategy in this paper are verified.

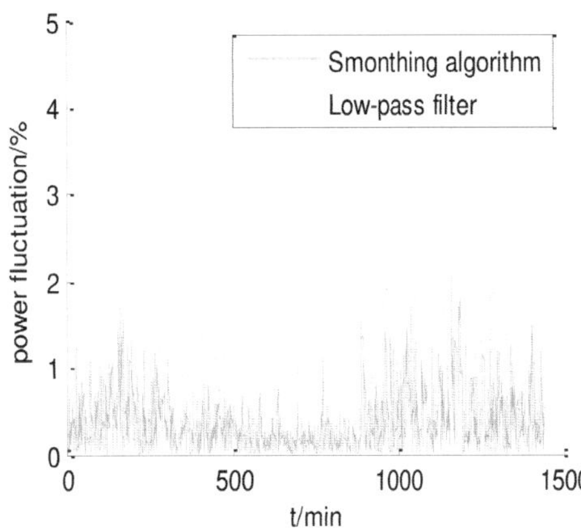

Figure 5. Comparison of wind power fluctuation rate of one minute

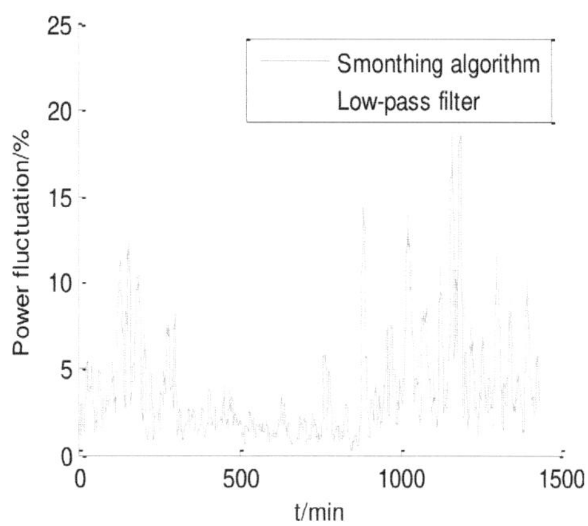

Figure 6. Comparison of wind power fluctuation rate of ten minutes

The parameters of fixed fluctuation coefficient and the algorithm are compared in Table 2. In order to make the wind power grid-connected power meet the fluctuation requirements of one minute and ten minutes, when the fixed fluctuation coefficient is 0.1, the fluctuation requirements of the two time scales are basically satisfied. The required rated power is 2.5501 MW and the required capacity is 3.1192 MW.h. By using the algorithm presented in this paper, the required power rating is 2.2347 MW and The required capacity of the required configuration is 2.3359 MW.h. Although the volatility increased slightly in ten minutes, it fully meets the wind power grid connection requirements.

Compared with the fixed fluctuation coefficient, the control strategy of the algorithm in this paper greatly reduces the charging and discharging amount of the energy storage device during sampling time, reduces the energy storage capacity of the energy storage device, increases the use time of the energy storage device, and reduces the initial cost and maintenance cost of the system.

Table 2. Comparison of fixed fluctuation coefficient

Fluctuation coefficient	Total charge and discharge (MW)	Rated power (MW)	Rated capacity (MW.h)	One minute maximum volatility	ten minutes maximum volatility
0	43.6344	4.7462	70.6925	0	0
0.1	10.5407	2.5501	3.1192	2.1%	12.4%
0.2	7.6370	1.6323	1.7401	3.4%	18.9%
0.3	6.3499	1.5389	1.5389	4.9%	23.2%
0.4	5．3152	1.1786	1.4731	6.5%	26.1%
0.5	4.3755	0.9765	1.3754	8.2%	28.3%
0.6	3.4985	0.7772	1.3340	9.7%	29.5%
0.7	2.6423	0.5770	1.3043	11.2%	30.2%
0.8	1.7853	0.3773	1.2818	12．6%	30.6%
0.9	0.9083	0.1830	1.2642	13.7%	30.9%
1	0	0	0	14.9%	31.1%
Algorithm	9.5144	2.2347	2.3359	2.0%	18.6%

6 Conclusion

Based on the study of the smooth wind power fluctuation of energy storage system, this paper proposes a wind power smoothing strategy based on adaptive adjustment of fluctuation coefficient. The simulation analysis of the actual wind farm data is used to verify the feasibility of the method.

(1) Particle swarm optimization algorithm based on sliding window can further improve the wind power smoothing effect. In comparison with the first-order low-pass filtering algorithm, we can find that the fluctuation of the grid-connection power of one minute and ten minutes has been significantly reduced.

(2) The fluctuation coefficient is optimized by the particle swarm algorithm of sliding window, so that the fluctuation coefficient is adjusted adaptively. In the case that the wind power grid-connected fluctuation is satisfied, compared with the fixed fluctuation coefficient, the total amount of energy storage charge-discharge during the controlling is reduced, so as to reduce the number of charge-discharge times of the energy storage system in the wind power grid-connected process, which is helpful to extend the life of the energy storage device and reduce the storage capacity of the system and the cost of system.

This control strategy can not only process historical data, obtain small energy storage capacity and reduce system cost, but also stabilize the grid-connected wind power, so as to reduce the phenomenon of 'abandoned wind power'. In the follow-up work, further consideration should be given to the control mode of energy storage system in order to achieve better wind level suppression effect.

7 References

[1] Liu C, Shi W H, Zhao Hai X 2014 Comparison of china's wind power integration standard with similar foreign standards *Smart Grid* 2 48-51.

[2] Zheng C 2011 Analysis of wind energy resources in global sea areas *Sino-global Energy* 16 37-41.

[3] Zhang T 2015 Research on real-time control strategy and capacity determination for hybrid

energy storage in mitigating wind power fluctuations .*Zhejiang University*.
[4] Su Q Q 2015 Research on the smoothing control of the battery energy storage system in wind-solar-battery hybrid power system *Shanghai Dianji University*.
[5] Jiang Q, Wang H 2013 Two-time-scale coordination control for a battery energy storage system to mitigate wind power fluctuations *IEEE Transactions on Energy Conversion* 28 52-61.
[6] Wang J Q 2016 Research on wind power to stabilize fluctuations in hybrid energy storage system *Northeast Petroleum University*.
[7] Li X 2012 Fuzzy adaptive Kalman filter for wind power output smoothing with battery energy storage system *IET Renewable Power Generation* 6 340-347.
[8] Zhang F, Meng K, Xu Z, et al 2017 Battery ESS planning for wind smoothing via variable-interval reference modulation and self-adaptive SOC control strategy *IEEE Transactions on Sustainable Energy* 8 695-707.
[9] Wang Z Y, Ma W, Hu R N 2017 Power Fluctuation stabilization control strategy of micro grid *Acta Energiae Solaris Sinica* 38 2344-2348
[10] Wang W S, Chi Y N, Zhao Z K, et al 2016 Standard on connecting wind farms to power system *China Standardization* 77 86-89.
[11] Yuan G L, Liu Y, Chen S L 2015 Smoothing control strategy of wind power based on dynamic fluctuation coefficient *Advances in New and Renewable Energy* 3 325-330.
[12] Zhou Y, Yan Z, Li N H, et al 2017 Design of multiple soc feedback starting in wind power smoothing algorithm using bess. *Acta Energiae Solaris Sinica* 38 1459-1467.
[13] Wang H J, Jiang Q Y 2014 An overview of control and configuration of energy storage system used for wind power fluctuation mitigation *Automation of Electric Power Systems* 38 126-135.

IOP Conf. Series: Earth and Environmental Science **342** (2019) 012025 doi:10.1088/1755-1315/342/1/012025

Method and development trend of corrosion state prediction of grounding grid

H Huang[1], Y C Liu [2,*], Y Xing[1], J R Wu[1], X Y Mao[1], X H Ma[1] and H B Sun[2]

[1]Electric Power Research Institute of Guizhou Power Grid Co., Ltd.,Guiyang,China 550002
[2]College of Electrical and Information Engineering,Hunan University,Changsha,China

Email: 491276835@qq.com

Abstract. With the development of modern power grids to ultra-high voltage, large capacity and long-distance, the grounding short-circuit current of power systems is getting larger and larger, and higher requirements are put forward for the safe and stable operation of grounding systems.The importance of studying the corrosion status of grounding grids is even more prominent. In this paper, from both static and dynamic aspects, the corrosion state prediction methods of the grounding grid are discussed. And the prediction technology is prospected, it is pointed out that the development of a device combining detection and prediction techniques will fill the gap in the international arena. It can not only provide effective guidance for corrosion protection of grounding grids, but also can be applied to other fields.

1. Introduction

Since the grounding grid is in the underground environment for a long time, after a certain period of operation, the conductors and grounding leads of the grounding grid will have different degrees of corrosion, and even break when severe.As a result, the cross section of the grounding body is reduced, the thermal stability is insufficient, and the electrical performance is deteriorated.At this time, in the event of a ground fault, a large safety accident may occur, resulting in immeasurable economic loss and adverse social impact.

At present, some phase results have been achieved in the corrosion detection of grounding grids at home and abroad[1-3], but even if the conductors are found to be corroded and thinned, it is impossible to know when the conductors no longer meet the requirements for the flow. Therefore, in order to grasp the operation of the grounding grid in time,and to provide early warning and maintenance of the equipment, it is necessary to conduct research on the prediction method of the corrosion state of the grounding grid. So far, there have been some theoretical research results on the prediction of corrosion state of grounding grids, but there are no reports on the results of corresponding equipment or instruments.

With the development of modern power grids to ultra-high voltage, large capacity and long-distance, the requirements for grounding systems are becoming higher and higher, and the soil corrosion problem of the grounding grid can not be ignored. Therefore, the importance of the corrosion prediction of the grounding grid is more prominent. It can effectively avoid the occurrence of safety accidents and has important theoretical and practical significance.

According to the different input physical quantities of corrosion rate prediction, the prediction method of corrosion state of grounding grid is introduced from two aspects in this paper. On the one hand, the factors affecting the corrosion rate of grounding grids are analyzed and the static prediction

IOP Conf. Series: Earth and Environmental Science **342** (2019) 012025 doi:10.1088/1755-1315/342/1/012025

method is discussed. On the other hand, the dynamic prediction method is introduced based on the historical corrosion rate data. Finally, suggestions and prospects of the corrosion prediction technology are put forward according to the current situation of grounding grid.

2. Static prediction of corrosion state of grounding grid

Since the grounding grid is generally buried underground, the conductors that make up the grounding grid may be corroded by the soil, and soil corrosion factors will directly affect the corrosion state of the grounding grid.The main input for static prediction is soil corrosion.

2.1. Soil corrosion factors

Soil is a very complex system which is composed of three phases of solid, liquid and gas. Its properties and structure are heterogeneous and versatile.When the metal is in contact with different soils, different interface potentials are formed between the metal and the soil cross section, and a loop is formed through the soil, then the corroded battery is formed[4]. Therefore, all factors affecting the physical properties, chemical properties, electrochemical properties and macroscopic environment of the soil will have an impact on soil corrosion.

Studies have shown that there are many factors affecting the corrosion of the grounding grid in the soil[5-6], mainly water content, electrical resistivity and salt content.

The corrosion that occurs when the grounding grid is buried in the soil is mainly electrochemical corrosion. Moisture is a prerequisite for causing the soil to become an electrolyte and causing electrochemical corrosion.Therefore, the water content in the soil has a great influence on corrosion. In general, when the soil moisture is not saturated, the more water content, the faster the corrosion rate of the metal.

Soil resistivity is an index of soil conductivity. In most cases, soils with low resistivity are highly corrosive, and soils with high resistivity are weak, that is, soil corrosivity is inversely proportional to soil resistivity.

Salt is the main component of electrolyte, so the conductivity of soil is obviously affected by the soluble salts and their composition. In general, the conductivity increases with the increase of soluble salts, that is, the more soluble salts, the greater the corrosiveness of macroscopic cell corrosion.

In addition, soil porosity, pH, soil type, various anions and cations, redox potential, organic matter content and microbial bacterial activity also affect the corrosion of the grounding grid. These influencing factors are both interconnected and act together to accelerate the corrosion of the grounding grid.

Therefore, it is necessary to consider the corrosion condition of the soil where the grounding grid is located and study the interaction of soil corrosion factors, so as to predict the soil corrosion of the grounding grid and the corresponding anti-corrosion measures can be taken in time.

2.2. Static prediction

PanakajK.Sen, a scholar at Colorado State University, based on the test data of 44 different soil corrosions in the 12 years of the National Bureau of Standards, proposed an empirical formula for calculating the corrosion rate of steel in soil[7],as in equation (1),

$$M = 3.36 - 9.63(10^{-5})(P_1) + 0.29(P_2) + 0.34(P_3) + 0.012(P_4) \quad (1)$$

where M represents the corrosion rate (mm/a), P_1 represents the soil resistivity (Ω·m), P_2 represents the pH value, P_3 represents the water content (%), and P_4 represents the gas content (%).

According to the specific situation of China, considering the factors of economy and other aspects, most of the grounding grid materials in China use ordinary carbon steel or galvanized steel, which is more corrosive than copper or copper clad steel of foreign grounding grid.

In response to actual needs, Xi'an University of Science and Technology explored the static prediction method of corrosion state of grounding grid. Wang Shaobo selected 13 soil corrosion factors of the grounding grid as indicators. After using the principal component analysis method to calculate the correlation coefficient between each index and the corrosion rate, the principal component samples were subjected to multiple regression prediction, but the results were not satisfactory. Subsequently, the BP neural network prediction model was proposed. From the experimental results, the expected accuracy requirements were not met[8].

Based on the data of soil corrosion factors in Shanxi Province, Li Na established three different corrosion rate prediction models of the grounding grid through BP neural network progressive optimization algorithm, support vector regression machine and fuzzy analogy method.Among them, the model based on fuzzy analogy method has better prediction effect[9].

Since there are many factors affecting the corrosion of the grounding grid, it is difficult to achieve the desired effect by using a single method for prediction. Du Jingyi proposed "a new method for predicting the corrosion rate of grounding grids"[10]. The prediction process is shown in Figure 1. The corrosion prediction rate of the grounding grid is predicted by the comprehensive prediction method based on vector similarity and Support Vector Regression(SVR).Firstly, the soil corrosion factor is regarded as the feature vector and dimensionless processing. Secondly, the similarity between each measured sample vector and the feature vector is calculated. Finally, the corrosion rate of the grounding grid is predicted according to the similarity.

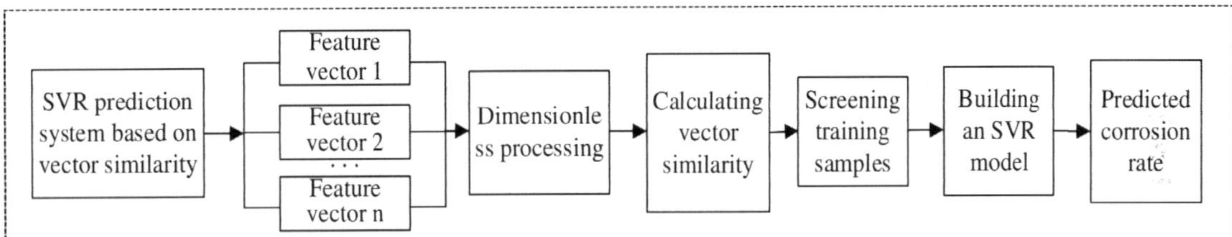

Figure 1. SVR corrosion rate prediction flow chart based on vector similarity.

3. Dynamic prediction of corrosion state of grounding grid

When the dynamic prediction method is used to predict the corrosion state of the grounding grid,the input is generally the historical corrosion rate data, and the output is the predicted value of the corrosion rate.

3.1. Historical measurement data

At present, according to the different principles, three major methods, such as electric network analysis method, electromagnetic field detection method and electrochemical method, are proposed to detect the corrosion of the grounding grid.The electrical network analysis method judges the corrosion of the grounding grid by measuring the magnitude of the grounding resistance. The electromagnetic field detection method uses the measured electric field or magnetic field distribution of the grounding grid and combines the numerical simulation of the grounding network to perform corrosion diagnosis. Electrochemical methods require the application of corrosion electrochemical detection techniques. Three methods have their own advantages and disadvantages.Therefore,the corrosion detection of the grounding grid is mainly realized by measuring the grounding resistance, detecting the grounding network connectivity and combining local excavation in the actual project, and the detection period is long and the sample data amount is small. These sample data can reflect the change of corrosion state in a certain period of time, so they have certain reference value.

Static prediction can not make full use of historical measurement data, but dynamic prediction can dynamically track historical data and reflect the trend of state change in time. Therefore, it is of certain significance to study the dynamic prediction of corrosion state of grounding grid, but it needs to choose appropriate method to deal with the nonlinear data of small samples.

3.2. Dynamic Prediction

The methods applied to corrosion rate prediction include neural network, gray model, support vector machine and combined prediction model of various methods. Each prediction model has its own advantages and disadvantages.

Hunan University has carried out some explorations to realize the dynamic prediction of the corrosion state of the grounding grid. In [11], BP neural network combined with grey prediction theory was used,and the combined forecasting model is constructed as shown in Figure 2.The corrosion rate of the first 3 months was used as input, and the output was predicted as the corrosion rate of the 4th month, and the corrosion rate of the next 9 months was derived. Due to the lack of sample data of the corrosion rate of the grounding grid, the prediction result of the combined model is not very satisfactory, and the longer the prediction time, the larger the error.The neural network method is used for fitting, and all historical data is used for training for each execution, thereby realizing dynamic prediction, but it does not really realize tracking of the innovation data.Because its prediction model will over-fitting the old data, it can't reflect the change of the innovation data in time.Kalman filter algorithm is an efficient algorithm for dynamic tracking of innovation data. It can design the dominant mode of tracking based on prior data information.Jing Jing applies Kalman filter algorithm to the dynamic tracking of corrosion state of grounding grid[12]. It avoids the drawbacks that the traditional corrosion state fitting algorithm can not achieve the tracking of new interest, and can better reflect the latest changes in corrosion state.

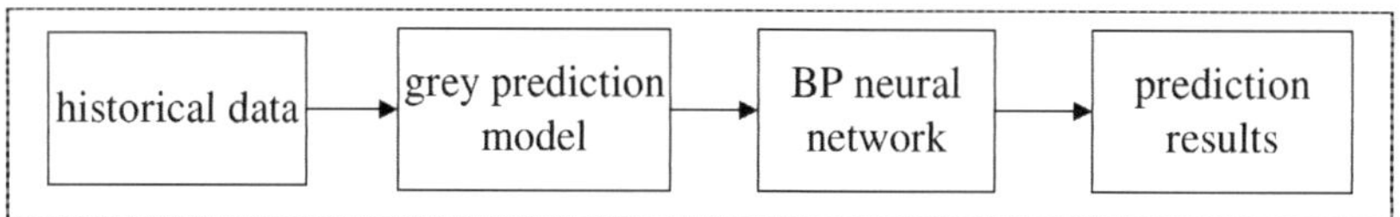

Figure 2. Combined forecasting model.

In general, the research results on the prediction of corrosion state of the grounding grid are still few, and the existing prediction methods also have some shortcomings.It is also necessary to build a model with a higher prediction accuracy.

4. Conclusion

The corrosion prediction study of the grounding grid plays an important role in the protection and modification of the grounding grid. While saving a lot of manpower, material and financial resources, it can prevent accidents and ensure the safe and stable operation of power system.

In the existing research, both static prediction and dynamic prediction methods have their own advantages and disadvantages, and the applicable conditions are different. The prediction accuracy of existing results still needs to be improved, and no corresponding instruments or equipment are developed.

With the development of modern power grids to ultra-high voltage, large capacity and long-distance, the short-circuit current of the power system is getting larger and larger, which puts higher requirements on the safe and stable operation of the grounding system. It is not difficult to foresee that the combination of corrosion state detection and prediction technology of the grounding grid is a future development trend. It is necessary to develop a corrosion state detection and prediction intelligent system of the grounding grid integrating data acquisition and detection analysis. It can quantitatively evaluate the corrosion state of the grounding grid and predict the development law.

Such a system can be applied not only to the grounding grid, but also to the grounding device of radar, communication and other systems. It can also be extended to corrosion detection and prediction of other underground metal facilities such as water supply, oil transportation, chemical pipelines, etc.It will have significant socioeconomic benefits.

Acknowledgment
This work is supported by the National Natural Science Foundation of China (No. 61472128, 61173108).

References

[1] Liu Y G,Meng X,Tian J H and Wang S 2010 Analysis on the method of selecting nodes on the erosion diagnosis of grounding grid *Journal of Chongqing University*) **33** 73-77
[2] Liu Y, Cui X, Zhao Z B, Li L and Chen J J 2009 Design and Application of Magnetic Field Detection System for Corrosion Diagnosis of Substation Grounding Grid *Journal of Electrotechnics* **24** 176-82
[3] Zhang X L, Luo P, Mo N,Wang Y G and Li Y L 2008 Development and application of electrochemical detection system for corrosion state of grounding grid *Chinese Society for Electrical Engineering* **28** 152-56
[4] Kong X Y 2005 Analysis of anti-corrosion of buried gas steel pipe *Shanxi Chemical Industry* **25** 61-63.
[5] Li S F, Chen Z Q, Peng M F and Yu D J 2003 Corrosion analysis and anticorrosion technology of grounding grid *Inner Mongolia Electric Power Technology* **21** 11-14
[6] Zhang C Z 1984 *Corrosion and Protection of Metals* (Beijing: Metallurgical Industry Press) pp 123-25
[7] Sen P K, Malmedal K and Nelson J P 2002 Steel grounding design guide and application notes *Int.Rural Electric Power Conference, 2002* (IEEE) pp C2-10
[8] Wang S B 2013 Application Research on Prediction of Corrosion Rate of Grounding Grid Based on Neural Network *Xi'an University of Science and Technology* pp 15-34
[9] Li N 2014 Research and prediction model of corrosion law of substation grounding grid *Xi'an University of Science and Technology* pp 32-52
[10] Du J Y, Du B Z, Han J and Yan A J 2015 A new method for predicting the corrosion rate of grounding grids *Mathematics Practice and Cognition* **45** 123-30
[11] Yang T 2011 Detection of corrosion state of grounding grid and its life prediction *Hunan University* pp 54-67
[12] Jing J 2016 Prediction method for corrosion state of grounding grid *Hunan University* pp 32-39

Research on data system framework and key technologies of EV charging station maintenance

Jiaqi Liao[1], Hualiang Fang[1], Kejun Qian[2], Chengke Zhoul[3], Luyang Dai[1]

[1]School of Electrical Engineering and Automation, Wuhan University, Wuhan 430072, China
[2]Suzhou Power Supply Company, Suzhou 215123, China
[3]School of Engineering and Built Environment, Glasgow Caledonian University, Glasgow, UK

Email: liaojiaqi@whu.edu.cn

Abstract. With the increase in the utilization rate of EV (Electric vehicle) charging facilities, the perception of operation status and the intelligent maintenance have become the key technologies of charging station development. A charging station data analysis method based on the two-dimension data framework is proposed according to the power transmission characteristic and data classification. Combined with the frequency and characteristic of fault events as well as the working principle of the charging station, the hardware and software framework based on the data processing and analysis for state evaluation and scientific maintenance planning is designed. Finally, the actual operation and maintenance statistics of the electric bus charging station in Suzhou, China is taken as an example for analysis.

1. Introduction

With more importance attached to environmental pollution and energy consumption, the state and local governments have introduced many policies to promote the electric vehicle industry. The operating environment of the charging facilities is mostly located outdoors with the environmental temperature and humidity uncontrollable, which results in the high failure rate. In addition, the lack of operation experience and technology of maintenance personnel result in the low maintenance rate and the long time of the equipment recovery. As the power grid terminal, the perception of the operation status and the scientific and intelligent maintenance have become the practical problems to be solved immediately.

So far, there are some researches on the operation and maintenance management methods of EV charging stations. Reference [1] has designed the charging facility monitoring system through system requirements analysis and functional service requirements, and divided the monitoring system into the system platform level, the supporting service level, and the public service level. Reference [2] proposed a set of embedded charging station power distribution monitoring system to realize data acquisition monitoring, analysis and evaluation, fault alarm and other functions. Reference [3] builds a monitoring system according to the AC and DC charging pile structure of different charging facilities. Reference [4] establishes charging station system model, device information model, communication model, and form a charging station monitoring system modeling program. Reference [5] has designed a system based on PCI bus with multi-DSP to collect the voltage and current of PCC to analyze the operation characteristic. The above researches mainly focus on the development and application of monitoring systems.

Therefore, based on the structure of the EV charging station, combined with the power transmission and operating characteristics of the charging station, considering the frequency and type of each fault event, starting from the working principle of the charging station, the hardware and software framework for state evaluation of the charging station and the scientific maintenance planning is proposed. Finally, the actual operation and maintenance statistics of the electric bus charging station in Suzhou is taken as an example for analysis.

2. Charging station structure

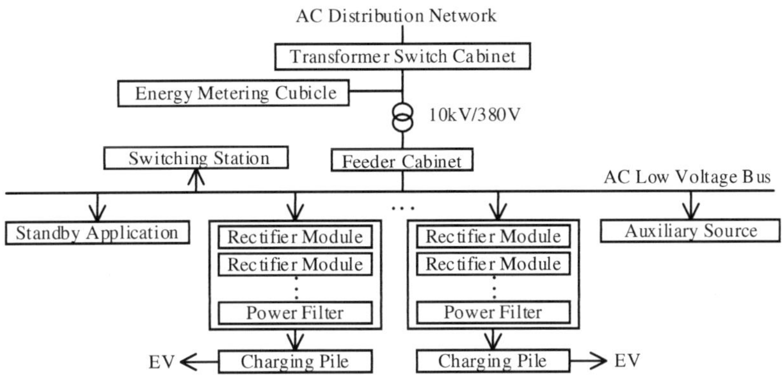

Figure 1. Structure of DC charging station

The DC charging station is set up as the city fast-charging station as well as the starting and terminal charging station of the bus, and the direct current is directly supplied to the electric vehicle. The DC charging power is higher and the charging time is relatively short. In addition, the DC charging station is under heavy load, and the rectifier cabinet is also a new type of equipment for the grid. As a result, the structure of the DC charging station is the basis of this study. The structure of DC charging station is shown in figure 1.

The power distribution process of the charging station is like to the traditional power distribution process. The main primary equipment includes transformer switch cabinet, main transformer and feeder cabinet. The secondary equipment includes high-voltage energy metering cubicle, relay protection device and control device. After each transformer, a wire cabinet is arranged, and a single busbar section wiring is arranged in the outlet cabinet, and a sectional switch is arranged between the busbars. Each feeder cabinet in the outlet cabinet distributes the electric energy to the charger in the rectifier cabinet and various auxiliary systems.

The charging cabinet in the rectification chamber is consisted of a plurality of charging modules in parallel according to different capacities and the core device of the charging station. The main body of the charger is connected to the feeder cabinet through a three-phase input contactor. Since a single power charging module is difficult to achieve high-power output and the reliability of the system needs to be improved, a distributed system must be implemented, that is, multiple identical power modules with the same current are connected in parallel. An active filter module is usually installed in the charger to reduce harmonics generated by the charging module that pollute the power grid.

The charging pile is the charging device that the EV user interacts with the power grid. The charging pile is internally provided with a TCU (Terminal Control Unit), a charging monitoring unit, and a charging execution unit. The charging monitoring unit can monitor the voltage, current, temperature of each key node in the charging station, the running status information of each switch device and communication device in the station. The charging execution unit can control the operation of the communication metering device such as the BMS (Battery Management System), the electric energy meter, the TCU, and the switching states of the switching devices in the station, as well as controlling the related devices according to the data input of the monitoring unit. The TCU can record the working information of the monitoring unit and the execution unit as well as the charging information and cost of the user using the charging pile.

3. Maintenance framework and fault events

Different from the traditional substation operation and maintenance method, the electric vehicle load of the charging station is intermittent, which is different from the ordinary continuous load in the power grid. In addition, as the interactive devices that non-professionals operate frequently, there are many potential failures for the charging piles. Besides, the distribution systems and rectification systems for all city fast charging stations are installed outdoors, which causes its operating status greatly affected by the season, weather and man-made.

3.1. Data analysis framework

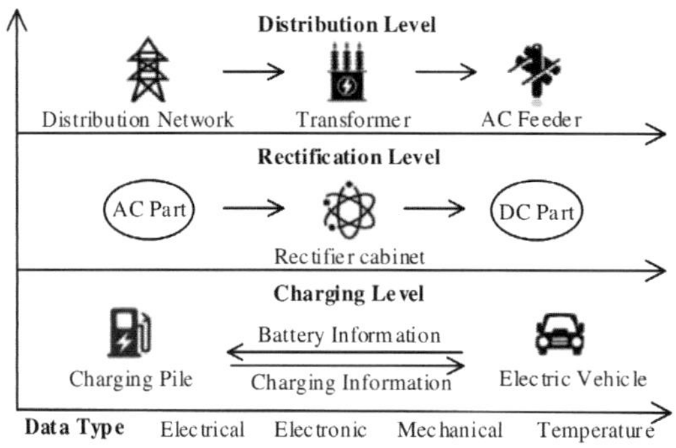

Figure 2. Two-dimension data system framework.

Combined with the structure of DC charging station, a two-dimension data analysis framework based on the status evaluation, fault analysis and maintenance are proposed to establish the data system of charging station. The transmission process of electric energy from top to bottom is divided into distribution level, rectification level and charging level. Each level is composed of electrical device, communication equipment, switch and others, which forms the vertical dimension. In addition, a large amount of data is generated while the equipment is running, which is closely related to the operating state of the charging station. According to the type and characteristics of data, it can be divided into electrical data, electronic data, mechanical data, temperature data, etc. The two-dimension data analysis framework is shown in figure 2. At present, the maintenance of electric vehicle charging stations still stay in the traditional stage with manual inspection. Although a large amount of historical operation data has generated, there is no specific data system to store and analyze the data. The main idea of the designed hardware and software is to make full use of historical data and real-time data through the two-dimension data analysis framework to realize the status monitoring, maintenance management and fault prediction of the charging station.

The main idea of the designed hardware and software is to make full use of historical data and real-time data through the two-dimension data analysis framework to realize the status monitoring, maintenance management and fault prediction of the charging station. The characteristic quantity is analyzed separately for each device in each level, and the fault information can be quickly located.

3.2. Fault events analysis

Various types of faults in the charging station are related to the characteristic quantities of the related equipment. For the distribution system level, it is like to the failure event analysis of a traditional substation. The main faults of electrical type include phase loss fault, short circuit fault, open circuit fault and bus switch status. Electronic faults mainly include state monitoring system faults, remote faults, etc. Main mechanical faults are abnormal wiring, joint damage, etc. The quantities of internal and external environment include three-phase winding temperature and humidity. By the horizontal dimension of the rectification system level, some common fault events can be summarized as shown in table 1.

Table 1. Fault events in rectification system level.

	Electrical	Electronic	Mechanical	Environment
AC Part	Undervoltage input, Undervoltage output, Contactor fault	Monitoring com-mnication fault, Measuring fault	Joint damage, Abnormal wiring	High temperature
Rectification	Charging module fault, Air switch trip	Module commu-nication fault	Loose connection, Abnormal wiring	High temperature, High humidity
DC Part	Under current input, Under current output	Measuring fault	Joint damage, Abnormal wiring	High temperature

For the charging system level, the interactive devices such as cables and display screens have various fault events, and there exists failures when the EV is changing information with charging piles as a result of human factors such as user operability and frequent operation. The partial fault events can be summarized as shown in table 2.

Table 2. Fault events in charging system level

	Electrical	Electronic	Mechanical	Other
Charging pile	Under current output, Undervoltage output, Charger insulation damage, Abnormal auxiliary power	Monitoring commni-cation fault, BMS communication fault, Abnormal data display, Offline	Cable damage, Card slot damage, Wiring fault, Screen damage	Cable high temperature, Arrester fault
Electric vehicle	Under current input, Battery damage	BMS handshake failure	Charging socket damage	

4. Hardware and software framework

4.1. Hardware framework design

For the hardware part, it consists of modules such as data input, central processing unit, upper management machine and data storage system, including data acquisition, data transmission and data storage. Firstly, the data is pre-processed through data cleaning, fusion and correction. Secondly, the data is layered and integrated to analyze for the state evaluation and fault diagnosis. In addition, the data platform is expanded by the network or cloud technology. Similarly, if the fault judgment module diagnoses that the equipment of charging station system is running under abnormal condition, the maintenance decision will be output. The hardware framework design of operation and maintenance system is shown in figure 3.

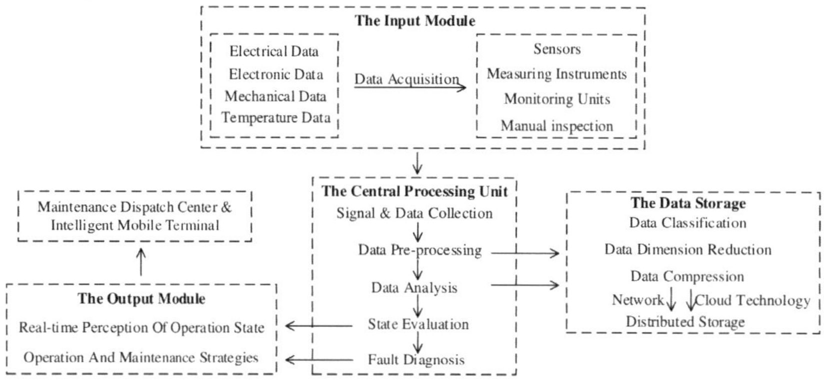

Figure 3. Hardware framework design of operation and maintenance system

For the collection of electrical data, the sensors such as voltage and current sensors are directly positioned. The switching data is analyzed by installing the vibration measuring instrument for vibration signal acquisition and the current sensor for signal processing analysis [6]. For communication data, it can be obtained from the process information downloaded from the monitoring board. The mechanical data can be periodically recorded by the manual inspection method. The temperature and humidity sensors can be directly installed for monitoring the internal and external environmental data of the device.

4.2. Software framework design

Figure 4. Software framework design of operation and maintenance system

For the software part, the state evaluation will perform data analysis on the data uploaded to the central processing unit after acquisition to estimate the status of the charging station and upload the result to the server in real time, so that the operation and maintenance personnel can monitor the status of charging station anytime and anywhere. For the fault analysis, the central processing unit analyzes the key characteristic quantities of the operational data to determine whether it has abnormal conditions, and quickly infers the type of fault events and locates them, as well as the warnings of potential faults and offline. For the fault event, the historical processing method is called from the event database, and the real-time maintenance strategy is formulated for reference to the operation and maintenance personnel. The improved software design framework of operation and maintenance system is shown in figure 4.

The software is composed of maintenance component level, common service level, and supporting service level. The maintenance component level includes the monitoring information visualization part and the basic application part for the maintenance personnel maintenance company. The operation monitoring includes the distribution system, rectification system and charging information. The common service level includes data processing and technology management. The function of technology management is information visualization generation, which includes maintenance strategy, system information, report output and network platform. The supporting service level includes basic services for software operations and various databases including real-time data, historical data and fault events.

5. Application analysis

At present, the fault information of DC charging station can be obtained through the reporting of TCU to the monitoring unit for automatic monitoring and the detecting of the maintenance personnel manually .According to the statistical analysis of the common equipment defects and the common TCU reporting faults of the Suzhou company from January to August 2018, firstly, the frequency of the common equipment defect is shown in figure 5.

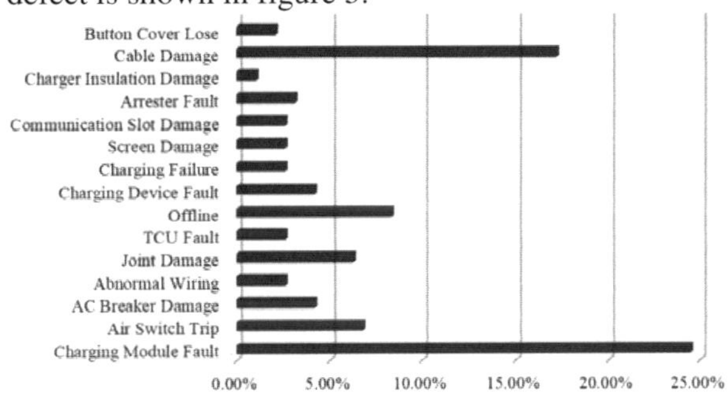

Figure 5. Frequency of common equipment defect

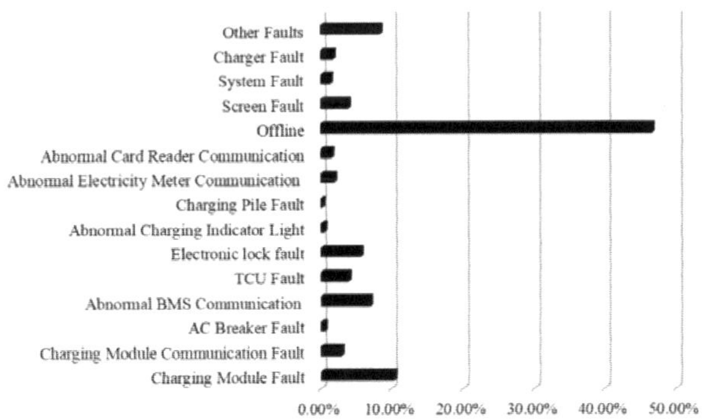

Figure 6. Frequency of common faults reported by TCU

It can be inferred that the most frequently occurring faults of the charging station during the current maintenance are charging module fault, cable damage and offline failure. Repairs can be made by directly replacing the module, replacing the fan unit, and changing the parameter data. The cable damage mainly includes the crack of charger and the repair method is to directly replace the cable. For the offline, the common way to solve the it is to restart the motherboard inside the charging pile. The maintenance framework studied in this paper contains monitoring and analysis methods for existing frequent faults, and forms operation and maintenance processing decisions based on fault characteristics.

The frequency of the common faults reported by TCU is shown in figure 6. For the faults that TCU reports, the offline fault is the most frequent, reaching 75% of the total fault. According to the TCU setting, after the offline fault occurs, the charging pile cannot be recorded, so the charging pile is in a fault state and is suspended. Other faults monitored by the TCU include the auxiliary power supply damage, the motherboard of the monitoring actuator unit damage, program not updated, audio cable damage, etc.

6. Conclusion

The research on hardware and software will eventually be popularly applied to the bus charging station that has been put into operation. Through the combination of the research and technology of this paper and the maintenance work of the charging station, it can visually monitor and evaluate the operating status of each charging facility, and make a quick judgment and positioning of the fault.

Subsequent research should consider the faults more comprehensively, and with the development of new energy sources, photovoltaic charging station and energy storage should also be considered. Further, the analysis method and application of artificial intelligences should be combined with the operation and maintenance to solve the issues of charging station and provide EV users great charging services.

Reference

[1] M. Zhao, G. Wang, Y. Wang and G. Sun 2011 Design and implementation of monitoring system for electric vehicle charging facilities Automation of Electric Power System **35(10)** p 65-69, in Chinese.

[2] J. Zhang, S. An and D. Gao 2018 Development of power distribution monitoring system for electric vehicle charging station based on Qt *Industrial Instrumentation & Automation* **2018.1** p96-100, in Chinese.

[3] H. Yan, G. Li, L. Zhao and B. Wu 2009 Design and implementation of monitoring system for electric vehicle charging station *Power System Technology* **33(12)** p15-19, in Chinese

[4] Y. Hu, Z. Guo, Q. Liu and M. Tang 2013 Modeling scheme of electric vehicle charging station monitoring system based on IEC 61970/61850 *Automation of Electric Power System* **37(2)** p91-96, in Chinese.

[5] X. Nie, J. Liu, L. Xuan, H. Liang, S. Pu, Q. Wang and N. Zhou 2013 Online monitoring and integrated analysis system for EV charging station *2013 IEEE PES Asia-Pacific Power and Energy Engineering Conference (APPEEC)*, Kowloon, 2013 p1-6.

[6] Y. Yang, Y. Guan, S. Chen, J. Wang and K. Zhao 2018 Mechanical fault diagnosis method for high voltage circuit breaker based on sound signal *Proceedings of the CSEE* **38(22)** p6730-6737, in Chinese.

Research on the factors affecting carbon emission of energy consumption in the Beijing-Tianjin-Hebei region

Y S Huang, Y Q Yang, M S Shi

Department of Economics and Management, North China Electric Power University, Hebei, 071003, China

Email: 15176261181@163.com

Abstract. Quantitative analysis of the factors affecting carbon emissions has important guiding significance for reducing regional carbon emissions in the region. Therefore, this paper first systematically analyzes the changes in the total carbon emissions of energy consumption in the Beijing-Tianjin-Hebei region. Then, the improved STIRPAT model is used to analyze the influencing factors of carbon emissions from energy consumption. Finally, the co-integration test and error correction model are used to study the long-term equilibrium relationship and short-term changes of carbon emission and its influencing factors (economic level, population size, technological level and energy structure) in the Beijing-Tianjin-Hebei region.

1. Introduction

The emission of greenhouse gases such as CO_2 has brought about a negative impact on the development of human society. Reducing carbon emissions will be a protracted war in the process of China's economic development. As long as the status of coal-based energy utilization has not been fundamentally changed, China's fossil energy will inevitably have greater demand and put more pressure on the environment [1]. As one of China's new development business circles, the Beijing-Tianjin-Hebei region plays an important role in China's development, but it also faces serious problems of excessive energy consumption and excessive carbon emissions. Therefore, it is of great significance to study the influencing factors of carbon emissions from energy consumption in the Beijing-Tianjin-Hebei region.

At present, many scholars have conducted many in-depth studies on the factors affecting carbon emissions. Shuai, et al. combined the STIRPAT model with the use of the panel and time-series data to analyze the impacts of population, affluence and technology on the carbon emission of 125 countries at different income levels over the period of 1990-2011 [2]. Fernándezamador, et al. used the ARDL boundary test to study the cointegration relationship between Pakistan's carbon emissions, energy consumption, trade openness and financial development [3]. Al-Mulali, et al. used the STIRPAT model to analyze the relationship between carbon emissions, urbanization levels and energy consumption in MENA [4]. Chikaraishi proposed an alternative concept and analytical framework using the improved STIRPAT model to determine at what level of urbanization can cause less carbon dioxide in human activities [5]. Lv, et al. used the STIRPAT model to study the effects of specific sectors in the industrial system on environmental pollution through ridge regression [6].

However, these studies are mainly concentrated on the analysis of the national level. There are few studies on the influencing factors of carbon emissions at the provincial level. So, this paper establishes an extended STIRPAT model to quantitatively analyze the influencing factors of energy consumption carbon emissions in the Beijing-Tianjin-Hebei region. It is expected that these recommendations can provide reference for energy conservation and emission reduction in the region.

2. Analysis of carbon emissions in the Beijing-Tianjin-Hebei region

2.1. Calculation of carbon emissions

This paper calculates the calculation according to the algorithm published in the 2006 IPCC guidelines. The specific accounting methods are as follows:

$$CE = \sum E_i \times NCV_i \times CEC_i \times COF_i \times \frac{44}{12} \quad (1)$$

Where CE is the total amount of emissions, E_i is the amount of energy consumed, NCV_i is the low calorific value of energy, CEC_i is the carbon content per unit of calorific value of energy, and COF_i is the carbon oxidation rate.

Considering the availability of data, only eight types of energy are considered in this paper. The related indicators of various energy sources are shown in table 1.

Table 1. Carbon emission factors and related indicators of energy sources.

Energy	Average low calorific value	Folded coal coefficient	Unit calorific value carbon content(ton/ TJ)	Carbon oxidation rate	Carbon dioxide emission coefficient
Raw coal	20908kJ/kg	0.7143kgce/kg	26.37	0.94	1.9003kg-co_2/kg
Coke	28435kJ/kg	0.9714kgce/kg	29.5	0.93	2.8604kg-co_2/kg
Crude oil	41816kJ/kg	1.4286kgce/kg	20.1	0.98	3.0202kg-co_2/kg
Gasoline	43070kJ/kg	1.4714kgce/kg	18.9	0.98	2.9251kg-co_2/kg
Kerosene	43070kJ/kg	1.4714kgce/kg	19.5	0.98	3.0179kg-co_2/kg
Diesel	42652kJ/kg	1.4571kgce/kg	20.2	0.98	3.0959kg-co_2/kg
Fuel oil	41816kJ/kg	1.4286kgce/kg	21.1	0.98	3.1705kg-co_2/kg
Natural gas	38931kJ/m^3	1.3300kgce/m^3	15.3	0.99	2.1622kg-co_2/m^3

2.2. Analysis of carbon emissions in the Beijing-Tianjin-Hebei region

According to the calculation method of carbon emissions, the total carbon emissions and energy consumption trends from 2000 to 2016 in the Beijing-Tianjin-Hebei region are shown in table 2.

Table 2. Total carbon emissions and energy consumption trends.

Year	Energy consumption (million tce)	Carbon emission (million tons)	GDP (billion RMB)	Coal share (%)	Carbon intensity (ton / million RMB)
2000	190.34	484.08	990.75	64.95	489
2001	195.31	497.76	1114.38	65.65	447
2002	208.70	534.25	1248.40	65.71	428

2003	231.50	595.16	1450.65	63.96	410
2004	265.74	685.14	1762.18	63.23	389
2005	313.23	810.90	2088.73	62.51	388
2006	333.16	864.21	2404.81	60.48	359
2007	339.96	878.19	2870.69	60.04	306
2008	371.21	960.64	3384.60	59.92	284
2009	393.42	1018.48	3691.04	60.46	276
2010	418.78	1079.90	4373.23	57.06	247
2011	454.70	1172.76	5297.50	56.40	221
2012	464.68	1200.53	5734.83	57.44	209
2013	466.60	1207.31	6268.58	58.47	193
2014	449.94	1161.99	6647.89	57.31	175
2015	449.64	1154.42	6935.89	56.34	166
2016	435.91	1118.81	7562.50	54.38	148

According to the changes in the total carbon emissions from 2000 to 2016, it can be seen that with the increase of GDP in the Beijing-Tianjin-Hebei region, carbon emissions show a significant growth trend. In 2000, the total carbon emissions in this region was 484.08 million tons, growing at an average growth rate of not less than 7% to 1118.81 million tons in 2016. Among them, from 2000 to 2013, carbon emissions have been growing at a relatively fast rate, and from 2013 to 2016, carbon emissions have gradually decreased. Although carbon emissions and energy consumption have been growing most of the time, the carbon intensity during these years has remained steadily decreasing.

3. Extended STIRPAT model

3.1. STIRPAT model

In 1972, Ehrlich and Holden first proposed the I=PAT equation to reflect the impact of human economic activity on environmental stress. In 1994, Dietz et al. established a stochastic model of the I=PAT equation-the STIRPAT model.

$$I_i = aP_i^b A_i^c T_i^d e_i \tag{2}$$

Where a is the model coefficient, b, c, d respectively represent the elastic coefficient of environmental impact, population size, average wealth and technical level, and e is the error term. Firstly, compared with the IPAT model, the STIRPAT model can introduce multiple independent variables. Secondly, STIRPAT is a nonlinear model. The introduction of the index makes the model can be used to analyze the non-equal effects of human factors on the environment.

3.2. Extended STIRPAT model

This paper expands the STIRPAT model to fit the carbon emissions. According to the environmental Kuznets curve theory, there is no linear relationship between economic development and carbon emissions, so it is replaced by quadratic. In the establishment of the extended STIRPAT model, in order to avoid zero or negative indicators, all independent variables are logarithmized. The extended STIRPAT model of this paper is expressed as follows:

$$\ln C = \ln a + \beta_1 \ln A^2 + \beta_2 \ln P + \beta_3 \ln T + \beta_4 \ln UR + \beta_5 \ln ES + \beta_6 \ln I \tag{3}$$

Where C is the carbon emissions, A is the average wealth, P is the population size, T is the technical level, UR is the urbanization level, ES is the energy structure, and I is the industrial level.

The specific values of the above indicators are shown in table 3.

Table 3. Specific data of indicators.

Year	Carbon emission (million tons)	Per capital GDP (RMB)	Number of resident population (million)	Carbon intensity (ton / million RMB)	urbanization rate (%)	Coal share (%)	Second industry ratio (%)
2000	484.08	10960.84	90.39	489	39.06	64.95	44.53
2001	497.76	12262.11	90.88	447	40.94	65.65	43.06
2002	534.25	13621.43	91.65	428	43.36	65.71	41.90
2003	595.16	15706.51	92.36	410	45.11	63.96	43.03
2004	685.14	18895.36	93.26	389	47.05	63.23	44.50
2005	810.90	22145.11	94.32	388	49.31	62.51	45.16
2006	864.21	25118.15	95.74	359	50.53	60.48	44.74
2007	878.19	29491.36	97.34	306	52.01	60.04	43.91
2008	960.64	34063.99	99.36	284	53.74	59.92	44.43
2009	1018.48	36465.48	101.22	276	55.48	60.46	42.82
2010	1079.90	41829.08	104.55	247	56.64	57.06	43.30
2011	1172.76	49057.91	106.15	221	57.78	56.40	43.80
2012	1200.53	53248.18	107.7	209	58.93	57.44	43.12
2013	1207.31	57404.55	109.2	193	59.66	58.47	42.03
2014	1161.99	60145.58	110.53	175	61.05	57.31	41.05
2015	1154.42	62244.36	111.43	166	62.51	56.34	38.40
2016	1118.81	67492.16	112.05	148	63.88	54.38	36.72

4. Analysis of factors affecting carbon emissions

4.1. Cointegration analysis

4.1.1. Unit root test. If the mean, variance, and autocovariance of the time series do not change over time, then the sequence is stable, otherwise it is unstable. If the stability test of the original data is ignored, the result may fall into the dilemma of pseudo-regression. Therefore, it is necessary to perform a stationarity test on the time series, and the unit root test is a method commonly used.

(1) Unit root test of sequence $\ln C_t$

The unit root test is performed on the sequence $\ln C_t$ by the ADF (Augmented Dickey-Filler) test, and the test results are shown in figure 1. It can be seen that the value of the t statistic is 0.777553, which is larger than the critical value when the significance level is 10%. Therefore, the original hypothesis cannot be rejected and the sequence is non-stationary.

		t-Statistic	Prob.*
Augmented Dickey-Fuller test statistic		0.777553	0.9991
Test critical values:	1% level	-4.667883	
	5% level	-3.733200	
	10% level	-3.310349	

Figure 1. ADF test results of sequence $\ln C_t$.

(2) Integration test of sequence $\ln C_t$

In order to determine whether the sequence $\ln C_t$ is monotonic, the unit root test should be performed on its differential sequence. The first-order difference sequence of $\ln C_t$ is denoted as $i\ln C_t$, and the results are shown in figure 2. The t-statistic value is -3.787347, which is less than the threshold value when the significance level is 10%, so the sequence $\ln C_t$ is a first-order single-sequence sequence, that is, $i\ln C_t \sim I(1)$.

		t-Statistic	Prob.*
Augmented Dickey-Fuller test statistic		-3.487347	0.0803
Test critical values:	1% level	-4.800080	
	5% level	-3.791172	
	10% level	-3.342253	

Figure 2. ADF test results of sequence $\ln C_t$.

(3) Unit root test of sequence $\ln A_t^2$, $\ln P_t$, $\ln UR_t$, $\ln T_t$, $\ln ES_t$ and $\ln I_t$

The ADF test is performed on sequences $\ln A_t^2$, $\ln P_t$, $\ln UR_t$, $\ln T_t$, $\ln ES_t$ and $\ln I_t$, and the test results are shown in table 4.

Table 4. Test results of sequences and their first-order and second-order difference sequences.

Sequence	T statistic	T statistic probability value	Conclusion
$\ln A_t^2$	0.586148	0.9984	unstable
$i\ln A_t^2$	-3.793613	0.0498	stable
$\ln P_t$	-3.499378	0.7570	unstable
$i\ln P_t$	-5.394587	0.0314	stable
$\ln UR_t$	-2.460117	0.3398	unstable
$i\ln UR_t$	-3.090888	0.1431	unstable
$ii\ln UR_t$	-6.244102	0.0011	stable
$\ln T_t$	-3.316561	0.1072	unstable
$i\ln T_t$	-4.679864	0.0121	stable
$\ln ES_t$	-2.098739	0.2480	unstable
$i\ln ES_t$	-3.499378	0.0234	stable
$\ln I_t$	0.242930	0.9958	unstable
$i\ln I_t$	-3.189930	0.1234	unstable
$ii\ln I_t$	-4.912810	0.0084	stable

The test results show that the t statistic probability values of sequences $\ln A_t^2$, $\ln P_t$, $\ln UR_t$,

IOP Conf. Series: Earth and Environmental Science **342** (2019) 012027 doi:10.1088/1755-1315/342/1/012027

$\ln T_t$, $\ln ES_t$ and $\ln I_t$ are greater than the test level of 10%, so that sequences $\ln A_t^2$, $\ln P_t$, $\ln UR_t$, $\ln T_t$, $\ln ES_t$ and $\ln I_t$ are considered to be unstable. While the t statistic probability values of the first-order difference se67quences $i\ln A_t^2$, $i\ln P_t$, $i\ln T_t$ and $i\ln ES_t$ of sequences $\ln A_t^2$, $\ln P_t$, $\ln T_t$ and $\ln ES_t$ are less than 10%, it is considered that the sequences $i\ln A_t^2$, $i\ln P_t$, $i\ln T_t$ and $i\ln ES_t$ are stable, that is, $i\ln A_t^2 \sim I(1)$, $ii\ln P_t \sim I(1)$, $i\ln T_t \sim I(1)$ and $i\ln ES_t \sim I(1)$. The t statistic probability values of the second order difference sequences $ii\ln UR_t$ and $ii\ln I_t$ of sequences $\ln UR_t$ and $\ln I_t$ are less than 10%, so that sequences $ii\ln UR_t$ and $ii\ln I_t$ are considered to be stable, that is, $ii\ln UR_t \sim I(2)$ and $ii\ln I_t \sim I(2)$. Therefore, there may be a long-term stable proportional relationship between the five groups of $\ln C_t$, $\ln A_t^2$, $\ln P_t$, $\ln T_t$ and $\ln ES_t$, and the next co-integration test is needed.

4.1.2. Cointegration test. If two or more variables are unstable, but some linear combination of them are stable, there is a long-term equilibrium relationship between these variables, namely the cointegration relationship. In this paper, the Johansen test is used to test whether there is a cointegration relationship between the five sequences $\ln C_t$, $\ln A_t^2$, $\ln P_t$, $\ln T_t$ and $\ln ES_t$.

Hypothesized No. of CE(s)	Eigenvalue	Trace Statistic	0.05 Critical Value	Prob.**
None *	0.992437	139.2541	60.06141	0.0000
At most 1 *	0.854554	65.98664	40.17493	0.0000
At most 2 *	0.723415	37.06735	24.27596	0.0007
At most 3 *	0.669775	17.78881	12.32090	0.0055
At most 4	0.074981	1.169114	4.129906	0.3259

Figure 3. Trace test results.

In figure 3, under the original hypothesis "At most 4", the calculated trace statistic value is 1.169114, which is less than the critical value of 4.129906. Therefore, it is acceptable to assume the null hypothesis that there are four cointegration relationships at the 5% significance level.

4.2. Error correction model

It has been proved above that there is a cointegration relationship between the five sequences $\ln C_t$, $\ln A_t^2$, $\ln P_t$, $\ln T_t$ and $\ln ES_t$. Therefore, an error correction model can be established to study the relationship between actual carbon emissions and economic level, population size, technical level and energy structure. The error correction model and related test results are shown in figure 4.

Variable	Coefficient	Std. Error	t-Statistic	Prob.
C	-16.09922	2.747019	-5.860613	0.0011
LNC(-1)	-0.566606	0.254306	-2.228051	0.0674
LNA2	0.511013	0.021243	24.05510	0.0000
LNA2(-1)	0.258143	0.129697	1.990353	0.0937
LNP	1.033811	0.238934	4.326755	0.0049
LNP(-1)	0.891859	0.361608	2.466370	0.0487
LNT	1.091137	0.036224	30.12195	0.0000
LNT(-1)	0.546814	0.248572	2.199825	0.0701
LNES	0.130283	0.080245	2.623574	0.0556
LNES(-1)	0.215060	0.091208	2.357922	0.0564

R-squared	0.999942	Mean dependent var	11.40571
Adjusted R-squared	0.999856	S.D. dependent var	0.299781
S.E. of regression	0.003603	Akaike info criterion	-8.145194
Sum squared resid	7.79E-05	Schwarz criterion	-7.662326
Log likelihood	75.16155	Hannan-Quinn criter.	-8.120467
F-statistic	11540.34	Durbin-Watson stat	2.130030
Prob(F-statistic)	0.000000		

Figure 4. ECM estimation and related test results.

The error correction model can be obtained as follows:

$$\begin{aligned}\ln C_t = &-16.09922 - 0.566606\ln C_{t-1} + 0.511013\ln A_t^2 + 0.258143\ln A_{t-1}^2 \\ &+ 1.033811\ln P_t + 0.891859\ln P_{t-1} + 1.091137\ln T_t + 0.546814\ln T_{t-1} \\ &+ 0.130283\ln ES_t + 0.215060\ln ES_{t-1} + \varepsilon_t\end{aligned} \tag{4}$$

The short-term changes in carbon emissions can be divided into two parts. The first part is due to changes in the current economic level, population size, technology level and energy structure. For each additional unit of $\ln A^2$ this year, the $\ln C$ of this year will change by 0.511013 units in the same direction, that is, the short-term elasticity of carbon emissions to the economic level is 0.511013. For each additional unit of $\ln P$ this year, the $\ln C$ of this year will change by 1.033811 units in the same direction, that is, the short-term elasticity of carbon emissions to the population size is 1.033811. For each additional unit of $\ln T$ this year, the $\ln C$ of this year will change by 1.091137 units in the same direction, that is, the short-term elasticity of carbon emissions to the technical level is 1.091137. For each additional unit of $\ln ES$ this year, the $\ln C$ of this year will change by 0.130283 units in the same direction, that is, the short-term elasticity of carbon emissions to the energy structure is 0.130283.

The second part is due to the changes in carbon emissions, economic level, population size, technology level and energy structure in the previous period. For each additional unit of $\ln C$ of the previous year, the $\ln C$ of this year will change by 0.566606 units in the opposite direction. For each additional unit of $\ln A^2$ of the previous year, the $\ln C$ of this year will change by 0.258413 units in the same direction. For each additional unit of $\ln P$ of the previous year, the $\ln C$ of this year will change by 0.891859 units in the same direction. For each additional unit of $\ln T$ of the previous year, the $\ln C$ of this year will change by 0.546814 units in the same direction. For each additional unit of $\ln ES$ of the previous year, the $\ln C$ of this year will change by 0.21506 units in the same direction.

5. Conclusion

This paper first analyzes the carbon emissions of energy consumption in the Beijing-Tianjin-Hebei region. Secondly, the improved STIRPAT model is used to analyze the influencing factors of carbon emissions from energy consumption. Then, using the cointegration and error correction model to study the long-term equilibrium relationship and short-term change between the carbon emissions and its influencing factors in Beijing-Tianjin-Hebei region. The results show that an increase in per capita GDP and population will increase carbon emissions, while the optimization of energy structure and technology will contribute to the reduction of carbon emissions. From the perspective of the impact on carbon emissions, the contribution of the technical level is the largest, the contribution of the economic level is second, and the contribution of the population size is the weakest.

Acknowledgments

This work was supported by the Energy Economic Development Strategy Research project of the Science and Technology Department of Hebei Province (grant number 12457206D-22).

References

[1] Du K, Xie C and Ouyang X. A comparison of carbon dioxide emission trends among provinces in China. 2017 *Renewable Sustainable Energy Rev.* **73** 19-25

[2] Shuai C, Shen L, Jiao L, Wu Y and Tan Y. Identifying key impact factors on carbon emission: evidences from panel and time-series data of 125 countries from 1990 to 2011. 2017 *Appl. Energy* **187** 310-25

[3] Fernándezamador O, Francois J F, Oberdabernig D A and Tomberger P. Carbon dioxide emissions and economic growth: an assessment based on production and consumption emission inventories. 2017 *Ecological Economics* **135** 269-79

[4] Al-Mulali U, Fereidouni H G, Lee J YM and Che N B C S. Exploring the relationship between urbanization, energy consumption, and CO_2 emission in MENA countries. 2013 *Renewable&Sustainable Energy Reviews* **23** 107-12

[5] Chikaraishi M, Fujiwara A, Kaneko S, Poumanyvong P, Komatsu S and Kalugin A. The moderating effects of urbanization on carbon dioxide emissions: A latent class modeling approach. 2015 *Technological Forecasting and Social Change* **90** 302-17

[6] Lv K J, Wang M Y and Wu T. Research on influence of Chinese industrial segment sectors on environmental pollution - Based on industrial network analysis. 2016 *Journal of Industrial Technological Economic* **35** 114-20

AUTHOR INDEX